Mohamed Ali El-Hodiri

Extrema of Smooth Functions

With Examples from Economic Theory

With 9 Figures

Springer-Verlag

Berlin Heidelberg New York
London Paris Tokyo
Hong Kong Barcelona
Budapest

Professor Dr. Mohamed Ali El-Hodiri
University of Kansas
Department of Economics
Summerfield Hall
Lawrence, KS 66045
USA

ISBN-13: 978-3-642-76795-1 e-ISBN-13: 978-3-642-76793-7
DOI: 10.1007/978-3-642-76793-7

2142/7130-543210 - Printed on acid - free paper

To My Parents
To My Teachers

INTRODUCTION

It is not an exaggeration to state that most problems dealt with in economic theory can be formulated as problems in optimization theory. This holds true for the paradigm of "behavioral" optimization in the pursuit of individual self interests and societally efficient resource allocation, as well as for equilibrium paradigms where existence and stability problems in dynamics can often be stated as "potential" problems in optimization. For this reason, books in mathematical economics and in mathematics for economists devote considerable attention to optimization theory. However, with very few exceptions, the reader who is interested in further study is left with the impression that there is no further place to go to and that what is in these second hand sources is all these is available as far as the subject of optimization theory is concerned. On the other hand the main results from mathematics are often carelessly stated or, more often than not, they do not get to be formally stated at all. Furthermore, it should be well understood that economic theory in general and, mathematical economics in particular, must be classified as special types of applied mathematics or, more precisely, of motivated mathematics since tools of mathematical analysis are used to prove theorems in an economics context in the manner in which probability theory may be classified. Hence, rigor and correct scholarship are of utmost importance and can not be subject to compromise.

It is our hope that this monograph well serve as a serious introduction to optimization theory which is selfcontained and where the theorems from mathematics are carefully stated, proved when the proofs are not inaccessible, and illustrated by numerical and economic theoretical examples. We have endeavored to provide an example of proper scholarship by citing original sources for the results from mathematics rather than using secondary sources. The clear benefits of this approach are that it minimizes the propagation of errors, that is makes the great body of high quality mathematical research available to economists, and that it represents an efficient resource allocation strategy by discouraging the re-inventions of the wheel and by encouraging specialization. Even though the process has been costly to us, we believe that the result justifies the cost. Indeed, we hope that the way economists apply mathematics may be changed for the better as a result of the example we tried to set here. One major objective is to lead the reader to consult primary sources first and then to devote more of the energy to the understanding of the economics which is after all what the mathematics is brought in to serve.

This monograph is written with three types of audiences in mind. It is directed at graduate students in economics who would like an introduction to optimization theory which takes them gradually as close as possible to the frontiers of the subject. It is directed to undergraduates in mathematics who would like to explore economics as an applied field in the social sciences. It is finally intended as a reference for professional economists where the results are clearly and formally stated and where references for further results are provided.

Even though the monograph is self contained, there are prerequisites which are covered by completing a course in mathematical analysis which uses a reasonably rigorous text book, e.g. Fleming (1977). For readers that are only familiar with multivariable calculus, the proofs may be difficult to follow but the statements of theorems and the examples are accessible.

This monograph represents an update of my earlier lecture notes, see El-Hodiri (1973), which has long been out of print. We have tried to improve the exposition and provided additional results, examples, graphs, references and corrections. Many of the mathematical results are not new but the presentation is. Some of the mathematics is new as we point out in places and reluctantly claim credit for them. For instance, the theorems on vector maximization in optimal control and the second order conditions for vector maxima in the finite dimensional case. Many of the economic applications are new and are claimed by the author e.g. the optimal tax and the optimal investment results which appear either here or in El-Hodiri (1973) for the first time.

The book is divided into two parts. Part I is devoted to finite dimensional problems. It starts gently with the unconstrained problem and moves on to deal with non-negativity constraints. This chapter is foundational in the sense that the problems presented in the remainder of the book can be, after a transformation, presented as applications of the results of that chapter. Indeed, that is one way of looking at the methods of proof adopted by us. Chapter 2 deals with equally constraints and is used to prove the theorems in chapter 3 which deals with inequalities as added side conditions. Chapter 4 is devoted to saddle value problems and to vector maxima. Many of the mathematical results are new as pointed out in the notes on the literature, but the main contribution of this part is that it presents a unified treatment of the subject, and it provides an example of a proper way to apply optimization theory to economic theory. Part II is devoted to variational and optimal control problems. Chapter 5 provides a fairly complete treatment of the unconstrained problems. Like chapter 1 it is foundational in the sense that its results are used to derive the results of the following chapters. Indeed, the results of chapter 5 are based on the results of chapter 1, of course after some work is done. Chapter 6 outlines the characterization of the problem of Bolza with equality constraints and its results are used to derive the results of chapter 6 where equalities are added constraints. Chapter 7 deals with optimal control problems and extensions as applications of the results of chapter 6. We have kept the survey chapter of El-Hodiri (1973) in its original form and it now appears as an appendix. This survey is well complemented by the monograph of Tikhomirov (1986), and by the outstanding survey article of Simon (1986).

In parts I and II we have relied heavily in Bliss (1946) which defines the spirit in which the current monograph is written. In view of the availability of such excellent treatment such as Pars (1940), Akheizer (1962), Berkovitz (1963), Gelfand and Fomin (1963), and Hestenes (1966) we have not provided as much detail in part II as there is in part I. Further and more advanced treatment of many of the topics presented in part I may be found in the books of Hestenes (1975) and of Ioffe and Tikhomirov (1974).

I was introduced to the proper methods of applying mathematics to economics by Professor Leonid Hurwicz whose influence is clearly present through out this work and I would like to take this opportunity to express my indebtedness to him for whatever is good in it. Of course I have evolved in bad ways and in other directions and he should be cleared from responsibility there. Professors James Quirk and Rubin Saposnik sat through at least one rendition of the material presented here and provided a lot of help to the author in terms of discussion, suggestion, and in terms of generosity of spirit the extent of which is rare among academics. My students at Purdue University and the University of Kansas have suffered through successive versions of these notes since 1965 and have contributed a great deal to their development. I would like to acknowledge the kind encouragement and support of professors Robert Bearse, Ronald Olsen, Anthony Redwood, and Fred Van Vleck of the University of Kansas. Clearly, none of the above mentioned is responsible for the errors in monograph.

Thanks are also due to Ms Marianne Bopp and Werner Muller of Springer-Verlag for their patient and competent assistance. Last but not least thanks to Sharon Gumm and Todd Kirkham and Adriana Sommerville for the expert typing and graphics that transformed my barely legible manuscript to a camera-ready final product.

Mohamed El-Hodiri
Lawrence, Kansas
April 1991

CONTENTS

PART 1

FINITE DIMENSIONAL PROBLEMS

CHAPTER 1

NO MAJOR CONSTRAINTS

The mathematical content of this chapter is well known, and can be found in any reasonable book on multivariable calculus, e.g. Fleming (1977). Nevertheless, the chapter serves two purposes. First, it presents a wish list, an ideal for which we address the smooth optimization problem comprehensively. We pose certain questions more easily asked now than later. Second, the chapter samples some of the applications of smooth optimization theory to economic analysis.

We will always assume, unless otherwise stated, that our functions have continuous second order partial derivatives in all their variables so they are extra smooth (!) and are, for short, labeled *C^2 functions*.

Consider a function $f: R^n \to R$ (read from R^n to R) given by $x \mapsto f(x)$, where R^n is the Euclidean space of n-tuples $x = (x^1, ..., x^n)$ with x^a being a real number, $a = 1, ..., n$ and where (as usual) when $n = 1$ then 1 is suppressed. We are assuming that the components of the gradient f_x and the elements of the Hessian f_{xx} of the function f exist and are continuous in x at all points $x \in R^n$. If we have to, we write $\bar{f}_x = (\bar{f}_1, \bar{f}_2, ..., \bar{f}_n)$ where $\bar{f}_i = \left.\frac{\partial f}{\partial x_i}\right|_{x=\bar{x}}$ and write $\bar{f}_{xx} = (\bar{f}_{ij})$, where $i,j = 1, ..., n$ and $\bar{f}_{ij} = \left.\frac{\partial^2 f}{\partial x_i \partial x_j}\right|_{x=\bar{x}}$. We say that a point $\bar{x}$ provides a *local maximum* for the function f if there exists a neighborhood $N(\bar{x}) = \{x \in R^n: |x - \bar{x}| < \varepsilon > 0\}$ such that:

$$f(x) \leq f(\bar{x}) \text{ for all } x \in N.$$

In this case we write:

$$\bar{x} = \textit{argu max } f \text{ on } N.$$

If $N = R^n$ then $\bar{x}$ is said to furnish a *global maximum* of f and we write

$$\bar{x} = \textit{argu max } f.$$

A global maximum is a local maximum but, of course, the converse is not true as figure 1 shows.

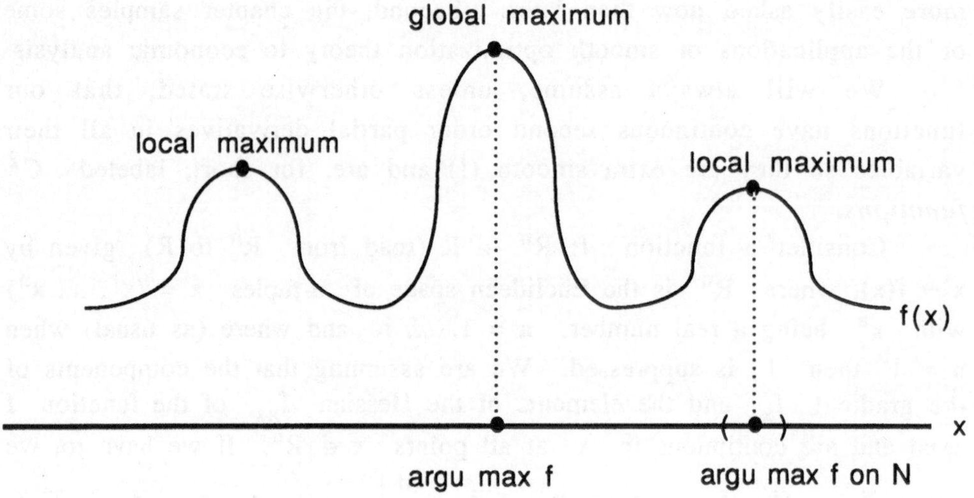

Figure 1. Local maxima and global maxima.

If the inequality defining the maximum is a strict inequality then we say that $\bar{x}$ provides a *strict maximum* of f of either type. If $\bar{x}$ provides a maximum of f then it provides a minimum of (- f) and so by having a theory of maximization we have a theory of extrema and that is why we will hardly talk about minima again, ever.

Exercise 1. Write down the definitions of local, global and strict minima. Find them on figure 1.

1. EXISTENCE

Investigating existence of solutions is almost an ethical requirement, especially in theoretical investigations (see the comment following equation 2 p. 78 in Evans (1924)). To plunge right into characterization of extrema would not only be ludicrous but could lead to fantastic results that are based on nothing whatsoever. An example cited by L. C. Young (see Young (1969)) provides a good illustration: If N is the greatest positive integer then $N = 1$. Indeed, if $N > 1$ then $N^2 > N$ and hence N is not the greatest integer □.

The easiest existence theorem in R^n is the *Weierstrass theorem* which states that:

A continuous function restricted to a closed and bounded subset A *of* R^n *attains its extreme values in* A.

Example: $f(x) = \frac{1}{x}$ on $[-1,1]$ attains neither maximum nor minimum there.

Exercise 2. Investigate the existence of extreme values of $f(x) = 2x$ on $(0,1)$ and of

$$g(x) = \begin{cases} \sin x & x \neq 0 \\ 2 & x = 0 \end{cases}$$

on R.

A better theorem, in the sense of not paying more for what could be obtained for less, states that:

If the function is upper semicontinuous and the set is closed and bounded then the function attains its maximum value on that set.

It should be noted that the above theorems provide sufficient but not necessarily necessary conditions for existence of extrema. With clearer conscience we march on.

2. FIRST ORDER NECESSARY CONDITIONS

These conditions state that the gradient of a function vanishes at a local maximum.

Let f be differentiable and let $\hat{x}$ = argu max f *on* N, *where* N *is an open subset of* R^n. *Then* $f_{\hat{x}} = 0$.

Exercise 3. Let $f(x) = -|x|$, $g(x) = 2x$ and let $h(x) = 1 - (x - 2)^2$. Consider the problem of maximizing f on R, g on R and h on $[0,1] = \{x \in R: 0 \le x \le 1\}$. Does the first order necessary condition hold for any of these functions. Why? Why not?

The point $\hat{x}$ where $f_{\hat{x}} = 0$ is called a *stationary point* or a *critical point* of f. The name comes from dynamics. Imagine a system whose motion is described by:

$$\frac{dx}{dt} = f_x.$$

Such a system, called a *gradient system*, will be in equilibrium at $\hat{x}$ if $f_{\hat{x}} = 0$. We lament the passing of the healthy tradition in economics writings of referring to first order necessary conditions as equilibrium conditions!

To prove the assertions that maximality implies stationarity, fix $h \in R^n$ and let $t \in R$ be small enough to assure that $(\hat{x} + th) \in N$. Then we have $f(th + \hat{x}) - f(\hat{x}) \le 0$. If $t > 0$ then

$\lim_{t \overset{+}{\to} 0} \frac{f(th + \hat{x}) - f(\hat{x})}{t} \geq 0$. If $t < 0$ then $\lim_{t \overset{-}{\to} 0} \frac{f(th + \hat{x}) - f(\hat{x})}{t} \leq 0$.
But since f is smooth the two limits coincide and we have:
$\lim_{t \to 0} \frac{f(th + \hat{x}) - f(\hat{x})}{t} = 0$. By continuity of the partial derivatives we have: the directional derivative in the direction of h is the same as the differential and $\hat{f}_x \cdot h = 0$. Since h is arbitrary, $\hat{f}_x = 0$. □

In the above proof it was essential that t could be positive or negative and that h is arbitrary. If for any reason we could not guarantee such freedom of movement then the theorem would have to be restated.

3. PROFIT MAXIMIZING FIRMS 1

First we consider a *price taking firm*, i.e. a firm which takes input and output prices as given. Let the profit function of the firm is given by

$$\pi(x,y\ ;\ p,q) = p \cdot y - q \cdot x,$$

where y is the firm's output and where x is the firm's input vector and where p and q are the firm's input and output prices respectively. Inputs and outputs are related by way of a twice differentiable production function, say, f so $y = f(x)$. Substituting in the profit function we get

$$\pi = p \cdot f(x) - q \cdot x.$$

If $\hat{x}$ is a local profit maximizing input vector then we must have:

$$p \cdot \hat{f}_x - q = 0$$

so that each input is paid the value of its marginal product.
If f is log linear, i. e. if

$$\ln f(x) = \ln \alpha_0 + \sum_{i=1}^{n} \alpha_i \ln x_i$$

then the first order necessary condition for profit maximization is

$$\alpha_i p\hat{y}/\hat{x}_i = q_i, \quad i = 1, ..., n$$

and the conditional demand for input i is given by

$$\hat{x}_i = \alpha_i p\hat{y}/q_i.$$

Thus a log linear technology implies that the profit maximizing input, in value terms is proportional to the value of the output.

Exercise 4. Evans (1930) Let the firm's cost function be given by: $\gamma(y) = Ay^2 + By + C$ and suppose the firm is a price taker. Show that the profit maximizing output is given by: $\hat{y} = (p - B)/2A$. Let the market demand be given by: $y = ap + b$. Show that the equilibrium price and output are: $p = (B + 2Ab)/1 - 2aA$ and $\hat{y} = (b + Ba)/1 - 2aA$.

Now consider a firm which is a monopolist, so that it faces an indirect demand function, say, given by: $y = \varphi(p)$. Let the cost function for this firm be denoted by $\gamma(y)$. The firm's profit function is: $\pi(y) = \varphi \cdot y - \gamma$. Let $\hat{y}$ be a profit maximizing output level. Then we have:

$$\hat{\varphi} + \hat{\varphi}' \cdot \hat{y} = \hat{\gamma}'$$

If we denote the demand elasticity by η, so $\eta = d \ln y/d \ln p$, then we can write the above equation as

$$p(1 + \frac{1}{\eta}) = MC$$

where MC = marginal cost = $\hat{\gamma}'$.

Exercise 5. Let $\gamma(y)$ be as in exercise 4 and let demand be given by: $y = ap + b$. Show that the maximizing output for the monopolist is given by:

$$\hat{y} = (b + Ba)/2 - 2aA.$$

4. FIRST ORDER SUFFICIENT CONDITIONS

If $\hat{x}$ is a critical point of a smooth function on f, then it may be argu max(f) or argu min(f) or none of the above. Indeed $\hat{x} = 0$ is a critical point of each of the functions: x^2, $-x^2$, and x^3, but is argu max for only one, one wonders which, on [-1,1].

If f *is concave and if* $\hat{x}$ *is a critical point of the smooth function* f *then* $\hat{x}$ = argu max(f) *globally*.

Before we prove our last assertion, we recall some facts about concave functions. To start with, a function $f: R^n \rightarrow R$ is said to be concave it every line segment connecting any two points on the graph of f never gets above that graph. In a less wordy way, let dom(f) be a convex set, f is concave if for any $\bar{x}$ and $\bar{\bar{x}}$ in dom(f) and for every $\theta \in [0,1]$ we have:

$$f(\tilde{x}) \geq \theta f(\bar{x}) + (1 - \theta) f(\bar{\bar{x}}).$$

where $\tilde{x} = \theta\bar{x} + (1 - \theta)\bar{\bar{x}}$.

The graph below illustrates the definition.

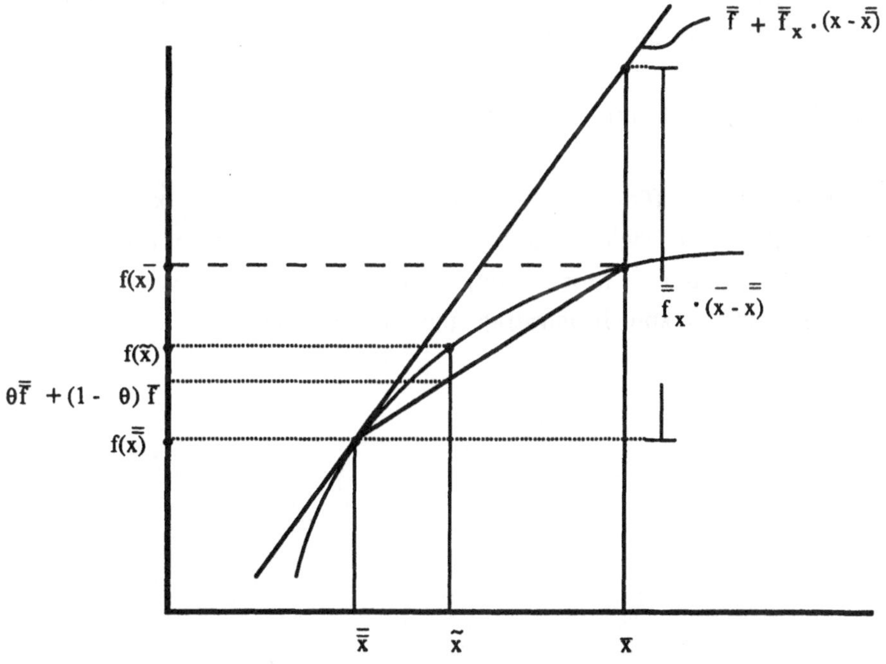

$\tilde{x} = \theta\bar{\bar{x}} + (1 - \theta)\bar{x}$.

Figure 2. A smooth concave function lies below its tangent line.

The tangent line of a smooth concave function lies on or above the graph of that function (see figure 2 above). This may be written as

$$f(\bar{x}) - f(\bar{\bar{x}}) \leq \bar{\bar{f}}_x \cdot (\bar{x} - \bar{\bar{x}}),$$

where f is smooth and concave and where $\bar{x}$ and $\bar{\bar{x}}$ are arbitrary points in dom(f).

Using the above inequality we can now prove our assertion on *first order sufficient condition.* Let $\hat{x}$ be a critical point of the smooth concave function f and let x be an arbitrary point in dom(f). Then we have:

$$f(x) - f(\hat{x}) \leq \hat{f}_x \cdot (x - \hat{x}) = 0. \square$$

<u>Exercise 6</u>. Noting that f is convex if and only if (-f) is concave, state and prove a first order sufficient condition for a minimum.

A question: Would quasi concavity do? Answer: No. A function is said to be *quasi concave* if all of its upper contour sets are convex sets i.e. $f: R^n \to R$ is quasi concave if the sets:

$$\{x \in R^n | \, f(x) \geq c\} \text{ are convex for all } c \in R.$$

Indeed, the function $f(x) = x^3$ is a quasi concave and $\hat{x} = 0$ is a stationary point of f which is not argu max f. More about quasi concave functions later. Meanwhile, temporary relief may be derived from the following graph illustrating the above example.

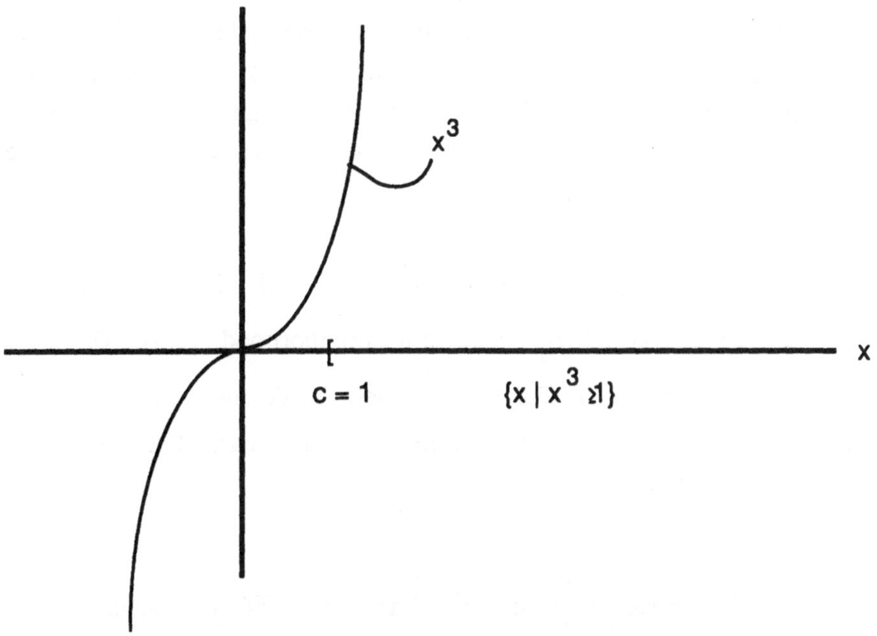

Figure 3. A quasi concave function.

5. SECOND ORDER NECESSARY CONDITIONS

Additional information can be derived from the fact that a point say $\hat{x}$ maximizes a C^2 function. The second differential of f at that point, given by $\hat{f}'' = a\hat{f}_{xx}a^*$, is non-positive. Such information is useful in conducting sensitivity analysis on either the theoretical or the computational level. Further, if we think of stationary points preliminary list of suspected maximizing points, then second order necessary conditions help to shorten the list by giving an alibi to some of the suspects -- if your second differential is positive then you have not done it --. The conditions are now stated.

If $\hat{x}$ = *argu max* f *on* N, *where* f *is* C^2 *and* N *is open, then:* i) $\hat{f}_x = 0$ and ii) $a\hat{f}_{xx}a^* \leq 0$ *for all* a *in* R^n, *where* a^* *is the transpose of* a *written as* $a^* = \begin{pmatrix} a^1 \\ a^2 \\ \vdots \\ a^n \end{pmatrix}$.

We have already proved (i) in section 4 and stating it here may not be cost effective since we acquired it earlier at a lower price .. we need f only to be differentiable. On the other hand, we are paying the higher price already so we might as well take delivery on the goods.

To prove (ii) we assume that, for some $\bar{a} \neq 0$, it is not true, i.e. $\bar{a}\hat{f}_{xx}\bar{a}^* > 0$. By continuity of f_{ij} there is a neighborhood S of 0 in R such that $\bar{a}\hat{f}_{xx}(\hat{x} + s\bar{a})\bar{a}^* > 0$ for all s there. We next apply Taylor's formula of order 2, i), and $\hat{x}$ = argu max f, taking care that the choice of neighborhoods is consistent to get:

$$0 \geq f(\hat{x} + s\bar{a}) - f(\hat{x}) = s\bar{a}\hat{f}_x + \frac{s^2}{2}\,\bar{a}\hat{f}_{xx}(\hat{x} + \tilde{s}\,\bar{a})\bar{a}^* = \frac{s^2}{2}\,\bar{a}\hat{f}_{xx}(\hat{x} + \tilde{s}\bar{a})\bar{a}^* > 0$$

The second order necessary condition used to be referred to as the stability condition which, of course, has its roots in dynamics. It turns out that a gradient system has an equilibrium at each of its stationary points as we pointed above. It also turns out that the linearized system is stable if and only if the second differential is non-positive. (see e.g. Beltrami (1987)).

6. SECOND ORDER SUFFICIENT CONDITIONS

If $\hat{x}$ *is a critical point of* f *which is* C^2 *and if second differential of* f *at* $\hat{x}$ *is negative for all nonzero increments then* $\hat{x}$ *is a point where* f *attains a strict local maximum.*

In symbols, if $\hat{f}_x = 0$ *and* $a\hat{f}_{xx}a^* < 0$ *for all* $a \neq 0$ *then* $\hat{x}$ = argu max f *locally and strictly.*

To prove the above theorem we first note that the assumption of continuity of the elements of f_{xx} at $\hat{x}$ and the assumption that

$a\hat{f}_{xx}a^* < 0$ for all $a \neq 0$ imply that there exists a neighborhood M of $\hat{x}$ such that $a\tilde{f}_{xx}a^* < 0$ for all $a \neq 0$ and all $\tilde{x} \in M$. Next we note that, using Taylor's formula of order 2 we may write:

$$f(\tilde{x}) - f(\hat{x}) = \frac{s^2}{2} \; af(\hat{x} + \bar{s}a)a^*$$

for some $\bar{s} \in (0,1)$.

Note that $\hat{x} + \bar{s}a \in M$ and hence $af_{xx}(\hat{x} + \bar{s}a)a^* < 0$. Thus $f(\tilde{x}) < f(\hat{x})$ for all $\tilde{x} \in M$. □

7. CONCAVITY, SECOND ORDER CONDITIONS, AND QUADRATIC FORMS

In the last 2 sections we dealt with the second differential of a function f of several variables. That second differential is a quadratic form generated by the Hessian of f, which under our assumptions that f is C^2, is symmetric. We say that a *quadratic form* $\xi A \xi^* = q(\xi)$ where A is a $n \times n$ matrix and where $\xi \in R^n$ is said to be *negative (positive) semidefinite* if $q(\xi) \leq 0$ (≥ 0) for all $\xi \in R^n$. When the inequality is strict and ξ is restricted to be nonzero then we say that $q(\xi)$ is *negative (positive) definite.*

Before we characterize semi-definiteness and definiteness, we introduce some terms from matrix algebra. Let A be $n \times n$ matrix. The *principal minors* of A of order s are the determinants of the submatrices of A given by

$$A_s = \begin{pmatrix} i_1 & i_2 & \cdots & i_s \\ i_1 & i_2 & \cdots & i_s \end{pmatrix}, \; i_1 < i_2 < \cdots < i_s,$$

where $\{i_1, i_2, ..., i_s\} \subset \{1, 2, ..., n\}$ and where the matrix A_s is given by

$$A_s = \begin{pmatrix} a_{i_1 i_1} & a_{i_1 i_2} & \dots & a_{i_1 i_s} \\ a_{i_2 i_1} & a_{i_2 i_2} & \dots & a_{i_2 i_s} \\ \dots & \dots & \dots & \dots \\ a_{i_s i_1} & a_{i_s i_2} & \dots & a_{i_s i_s} \end{pmatrix}.$$

The principal minor of order s which is the determinant of $\begin{pmatrix} 1 & 2 & \dots & s \\ 1 & 2 & \dots & s \end{pmatrix}$ is called the *principal diagonal minor* of order s of A.

Example. Let $A = \begin{pmatrix} 1 & 2 & 1 \\ 2 & 2 & 0 \\ 1 & 0 & 3 \end{pmatrix}$. Then the principal minors of order 1 are the determinants of (1), (2), (3). The principal minors of order 2 are the determinants of the matrices: $\begin{pmatrix} 1 & 2 \\ 1 & 2 \end{pmatrix}, \begin{pmatrix} 1 & 3 \\ 1 & 3 \end{pmatrix}, \begin{pmatrix} 2 & 3 \\ 2 & 3 \end{pmatrix}$ i.e. of the matrices $\begin{pmatrix} 1 & 2 \\ 2 & 2 \end{pmatrix}, \begin{pmatrix} 1 & 1 \\ 1 & 3 \end{pmatrix}, \begin{pmatrix} 2 & 0 \\ 2 & 3 \end{pmatrix}$. The principal minor of order 3 is the determinant of A. The successive principal diagonal minors of A are the determinants of (1), $\begin{pmatrix} 1 & 2 \\ 2 & 2 \end{pmatrix}$, A.

We state without proof the characterization of definiteness and semidefiniteness of quadratic forms of symmetric matrices (see e.g. Gantmacher (1959)). *A quadratic form* $\xi A \xi^*$ *is negative semidefinite if and only if all the principal minors of* A *of order* s *either are zero or have the sign of* $(-1)^s$, s = 1, ..., n. *It is positive semi definite if and only if all the principal minors of* A *of order* s *are non-negative for* s = 1, ..., n. *A quadratic form is negative definite if and only if all the principal diagonal minors of* A *of order* s *have the sign of* $(-1)^s$, s = 1, ..., n. *It is positive definite if and only if all the principal diagonal minors of* A *are positive.*

Example. Even though all the successive diagonal principal minors of $A = \begin{pmatrix} 0 & 0 \\ 0 & -1 \end{pmatrix}$ are non-negative, the quadratic for $\xi A \xi^* = -(\xi^2)^2$ is *not* positive semi-definite. It isn't negative definite either (take $\xi = (1\ 0)$).

We will revisit quadratic forms later. For now we are able to state second order conditions for local extrema of C^2 functions in terms of principal minors of the Hessian of f at the point being investigated. Exercise: Do that.

We move on to relating concavity and quasi concavity of C^2 functions to second order conditions. *A* C^2 *function is concave if and only if* $\xi f_{xx} \xi^* \leq 0$ *for all* $x \in \mathrm{Dom}(f)$ *and for all* ξ. In other words if and only if the second differential of f is negative semi-definite at all x where f is defined and C^2. If in the definition of concavity, the inequality is strict then we say that f is *strictly concave.* While it is true that the negative definiteness of the second differential at all points implies that the function is strictly concave, the converse is not true. Take $f: R \to R$ given by $f = -x^4$. If the restriction of f to an open subset M of dom(f) is (strictly) concave then we say that f is locally (strictly) concave.

We relate concavity to maximization of C^2 functions. We label some statements first. We always assume f is C^2.

1) $\hat{x}$ = argu max f locally (1' strictly)
2) $\hat{x}$ = argu max f globally (2' strictly)
3) $\hat{x}$ is a critical point of f
4) $\xi \hat{f}_{xx} \xi^* \leq 0$ for all ξ
5) $\xi f_{xx} \xi^* < 0$ for all $\xi \neq 0$
6) f is concave (6' strictly)
7) f is locally concave (7' strictly)

In the table below (read across) = means: "is equivalent", ✓ means: "the implication is valid", and × means "does not imply"

Table 1

	1	1'	2	2'	3	4	5	6	6'	7	7'
1	=	×	×	×	✓	✓	×	×	×	×	×
1'	✓	=	×	×	✓	✓	×	×	×	×	×
2	✓	×	=	×	✓	✓	×	×	×	×	×
2'	✓	✓	✓	=	✓	✓	×	×	×	×	×
3	×	×	×	×	=	×	×	×	×	×	×
4	×	×	×	×	×	=	×	×	×	×	×
3 and 5	✓	✓	×	×	✓	✓	✓	×	×	✓	✓
3 and 6	✓	×	✓	×	✓	✓	×	✓	×	✓	×
3 and 6'	✓	✓	✓	✓	✓	✓	×	✓	✓	✓	✓
3 and 7	✓	×	×	×	✓	✓	×	×	×	✓	×
3 and 7'	✓	✓	×	×	✓	✓	×	×	×	✓	✓

For every "×" in the table there is a counter example proving it.
<u>Exercise</u>: Find them (hint: see Roberts and Varberg (1973)).

8. PROFIT MAXIMIZING FIRMS 2

We can now prove that a firm's "supply curve" slopes up and that a "firm's input demand curve" slopes down. We do it here the hard way, for there is a much easier way, since we would like to show how the second order conditions are applied to theoretical sensitivity analysis. We use the notation of section 3. Let $\hat{x}$ = argu max $\pi(x)$. Drop the "hat" since we will vary p and q (input and output prices). x is a critical point of π and so

$$(\dagger) \quad pf_{x^i} - q^i = 0, \qquad i = 1, \ldots, n$$

Assume that f_{xx} is non-singular so that we can apply the implicit function theorem (see Fleming 1977) to solve for x as a function of p and q. We get the existence of a unique function $\xi(p,q)$ such that

$$(\dagger\dagger) \quad pf_{x^i}[\xi(p,q)] - q^i = 0$$

in a neighborhood of the initial value of (p,q). Assume further that $af_{xx}a^* < 0$ whenever $a \neq 0$. Then by the second order sufficient conditions the functions $\xi(p,q)$ maximize then profits of our price taking firm. Thus they deserve being called input demand functions. Substituting into $y = f(x)$ with $x = \xi(p,q)$ we get a unique function $y = \eta(p,q)$ which is entitled to the title of supply function of the firm. We will show that $\frac{\partial\eta}{\partial p} \geq 0$ and $\frac{\partial\xi^i}{\partial q} \leq 0$. Taking the total differential in $(\dagger\dagger)$ we have:

$$p\sum_j f_{ij}d\xi^i + f_{x^i}dp - dq^i = 0$$

multiply by $d\xi^i$ and sum over i to get

$$p\sum_i\sum_j f_{ij}d\xi^i d\xi^j + (\sum_i f_{x^i}d\xi^i)dp - \sum_i dq^i d\xi^i = 0$$

By the second order necessary condition and since $p > 0$ the first term is non-positive. Hence, recall $d\eta = \sum_i f_{x^i}d\xi^i$, we have

$$(\dagger\dagger\dagger) \qquad d\eta dp - dqd\xi^* \geq 0$$

Our assertion follows from $(\dagger\dagger\dagger)$.

9. NON-NEGATIVITY CONSTRAINTS

Variables may have to be constrained to be non-negative because of the "physics" of the model or because it is the way it has always been done. The problem then is to maximize a function $f: R^n \to R$ subject to the constraint that $x \geq 0$. We use the notation $x \geq y$ to mean $x^i \geq y^i$ for all i. Using a device introduced by Valentine (see Valentine (1937)) in a much more general contest we can transform the problem so that the theory developed in this chapter applies. The device replaces each x^i by $(z^i)^2$ as an argument of f. The problem of maximizing f(x) subject to $x \geq 0$ is equivalent to maximizing $g(z) = f((z^1)^2, (z^2)^2, ..., (z^n)^2)$ without restrictions on z. In more fancy terms we have transformed a problem maximization on a manifold with a boundary to a problem maximization on a manifold by working on the manifold's double (see Munkres (1968)). The *first order necessary condition for* $\hat{x}$ to be locally argu max f subject to $x \geq 0$, where f is C^2 is:

$$(1) \quad \hat{f}_x \leq 0, \qquad \hat{x}^i \hat{f}_{x^i} = 0.$$

To prove (1) we note that $\hat{z}$ maximizes g(z) if and only if $\hat{x}$ maximizes f(x) subject to $x \geq 0$. Since g is C^2 in z our first order necessary condition of section 2 apply. We get

$$\hat{g}_{z^i} = 2\hat{f}_{x^i}\hat{z}^i = 0$$

and since $\hat{z}^i = 0$ if and only if $\hat{x}^i = 0$ we get the second part of (1).

To prove the first part of (1) note that for $\hat{x}^i > 0$ we have $\hat{f}_{x^i} = 0$ by the second part of (1). So let I be the set of indices of components of $\hat{x}$ which are zeros, i.e. $\hat{x}^i = 0$ for all $i \in I$. If I is empty then we are done. If not we go on. The second order necessary condition of section 5 applies and hence $\sum_i \sum_j \hat{g}_{ij}\zeta^i\zeta^j \leq 0$ for all $\zeta \in R^n$. Hence in particular the inequality holds for a choice of ζ, call it $\bar{\zeta}$, where $\bar{\zeta}^i = 0$ for $i \notin I$, $1/\sqrt{2}$ otherwise.

Thus we have $\sum_{i\in I}\sum_{j\in I} \hat{g}_{ij}\bar{\zeta}^i\bar{\zeta}^j \le 0$. Now $\hat{g}_{ij} = 4\hat{f}_{ij}\hat{z}_i\hat{z}_j + 2\hat{f}_i = 2\hat{f}_i$ for $i \in I$. Hence $0 \ge \sum_{i\in I}\sum_{j\in I} \hat{g}_{ij}\bar{\zeta}^i\bar{\zeta}^j = \hat{f}_i$ □.

In the above proof we paid too much for the first order necessary condition since there are easy direct proofs that require only differentiability of f. On the other hand we, by adding a couple of lines, but no more assumptions, get a second order necessary condition for current problem. The second order necessary condition states that:

$$(2) \quad \sum_{i\in I} \hat{f}_i(\zeta^i)^2 + 2\sum_{i\notin I}\sum_{j\notin I} \hat{f}_{ij}(\hat{x}^i)^{1/2}(\hat{x}^j)^{1/2}\zeta^i\zeta^j \le 0, \quad \text{for all } \zeta.$$

If we, in addition to (1), assume concavity then $\hat{x}$ is globally argu max f over $x \ge 0$. If in addition to (1), we assume the strict inequality in (2) to hold for all $\zeta \ne 0$ then $\hat{x}$ is locally argu max f over $x \ge 0$.

Example. Maximizing $f(x,y) = -20x - x^2 - y^2$ subject to $x \ge 0$, $y \ge 0$. Clearly the maximum occurs at $\hat{x} = 0, \hat{y} = 0$. $\hat{f}_x = -20, \hat{f}_y = 0$ so the first order necessary condition is verified. And so are the second order conditions.

10. UTILITY MAXIMIZING CONSUMERS 1

Let $x \in R^n$ denote commodity bundles, let $p \in R^n$ denote a price vector taken as a parameter by the consumer, and let m be the liquid assets that the consumer devotes to consumption. We wish to avoid the confusing reference to m as income and we refer to m as liquidity. The consumer's utility is, here, given by a smooth function u(x) and the budget constraint is given by: $p \cdot x \le m$. The consumer's problem is to maximize u(x) subject to:

(1) $p \cdot x \le m, \qquad x \ge 0.$

We will assume that $p > 0$ and $m > 0$. The set of points x that satisfy (1) is given by:

(2) $$x^i = \frac{\alpha^i}{(\alpha^0,\alpha)\cdot(1,p)}\, m, \quad \text{where } 0 \le (\alpha_0, \alpha) \in R^{n+1}.$$

This can be verified by noting that

$$p \cdot x = \frac{p \cdot \alpha}{\alpha^0 + \alpha \cdot p}\, m \le m$$

with equality if $\alpha^0 = 0$. Thus α_0 represents unused liquidity (more on that later). Substituting into u from (2), the problem is to maximize:

$$u\left[\frac{\alpha}{(\alpha^0,\alpha)\cdot(1,p)}\, m\right], \qquad \text{where } \alpha \ge 0.$$

Let $(\hat{\alpha}^0,\hat{\alpha})$ solve the consumer's problem. Then the first order necessary conditions are:

(3.1) $$\sum_i \hat{u}_i \hat{x}_0^i \le 0, \qquad \hat{\alpha}_0 \cdot (\hat{u}_x)\hat{x}_0 = 0,$$

$$\text{where } \hat{x}_0^i = \left.\frac{\partial \hat{x}^i}{\partial \hat{\alpha}^0}\right|_{(\hat{\alpha}^0,\hat{\alpha})}, \quad \hat{u}_i = \left.\frac{\partial u}{\partial x^i}\right|_{(\hat{\alpha}^0,\hat{\alpha})}.$$

(3.2) $$\sum_i \hat{u}_i \hat{x}_j^i \le 0, \qquad \hat{\alpha}_j \cdot (\hat{u}_x)\hat{x}_j = 0,$$

$$\text{where } \hat{x}_j^i = \frac{\partial \hat{x}^i}{\partial \alpha^0}, \quad j = 1, \ldots, n.$$

Note that if $\hat{\alpha}^0 = 0$ then our consumer is at a point of satiation at $\hat{x}$. To see that, we observe that $x_0^i = \frac{-\alpha^i}{[(\alpha^0,\alpha)\cdot(1,p)]^2} m = \frac{-x^i}{(\alpha^0,\alpha)\cdot(1,p)}$.

Thus $u_i x_0^i = -u_i \frac{x^i}{(\alpha^0,\alpha)\cdot(1,p)}$. By (3.1) we, thus, have $\sum \hat{u}_i \hat{x}^i \geq 0$.

Assuming $\hat{u}_i \geq 0$ we conclude that $\hat{\alpha}^0 > 0$ implies $\hat{u}_i \hat{x}^i = 0$ for all i since $\sum \hat{u}_i \hat{x}^i = 0$. But then the consumer is satiated, otherwise there would exist an i_0 with $\hat{x}^{i_0} > 0$ and $\hat{u}_{i_0} > 0$. This observation and the unit "price" of α^0 supports interpreting m as liquidity (even more on this later). Next, let us consider $u_m = \frac{\partial u}{\partial m}$ which is the marginal utility of liquidity.

$$u_m = \sum u_i x_m^i = \sum u_i \frac{\alpha^i}{(\alpha^0,\alpha)\cdot(1,p)} = \frac{1}{m} \sum u_i x^i.$$

Our condition (3.1) is equivalent to:

$$\text{(3.1}')\qquad \hat{u}_m \geq 0, \qquad (m - p\cdot\hat{x})\cdot\hat{u}_m = 0.$$

We now work on (3.2). Note that:

$$x_j^i = \frac{\partial x^i}{\partial \alpha^j} = \begin{cases} \dfrac{[(\alpha^0,\alpha)\cdot(1,p) - \alpha^i p^i]}{[(\alpha^0,\alpha)\cdot(1,p)]^2} m, & i = j \\[2ex] \dfrac{-\alpha^i p^j m}{[(\alpha^0,\alpha)\cdot(1,p)]^2}, & i \neq j \end{cases}$$

In other words:

$$x_j^i = \begin{cases} \dfrac{m}{(\alpha^0,\alpha)\cdot(1,p)} - \dfrac{x^i p^i}{(\alpha^0,\alpha)\cdot(1,p)}, & i = j \\[2ex] \dfrac{-x^i p^j}{(\alpha^0,\alpha)\cdot(1,p)}, & i \neq j \end{cases}$$

Thus 3.2 can be written as:

$$\text{(3.2}')\qquad \frac{\hat{u}_i m}{(\hat{\alpha}^0,\hat{\alpha})\cdot(1,p)} - \frac{p^i \sum \hat{u}_j \hat{x}^j}{(\hat{\alpha}^0,\hat{\alpha})\cdot(1,p)} \leq 0, \text{ with equality if } \hat{x}^i > 0.$$

Multiply (3.2′) by $\frac{(\hat{\alpha}^0\alpha)\cdot(1,p)}{m}$ to get:

(3.2′′) $\hat{u}_i \leq \hat{u}_m p^i, \qquad \hat{x}^i\cdot(\hat{u}_i - \hat{u}_m p^i) = 0.$

Note again that, by 3.1′ and 3.2′′, if $p\hat{x} < m$ then $\hat{u}_i \leq 0$, $\hat{x}^i\hat{u}_i = 0$ which would be a bless point if u was concave. It says that if you got the money out of alternative uses and you didn't spend it all then you should see your economist. And soon.

Finally we look at the interior solution. Let $\hat{x} > 0$ be a solution to the consumer's problem and assume nonsatiation. Then

(4.1) $\hat{u}_i = \hat{u}_m p^i, \qquad i = 1, ..., n$

(4.2) $m = p\cdot\hat{x}.$

CHAPTER 2

EQUALITY CONSTRAINTS

By way of introducing the problem we deal with in this chapter, consider the problem of maximizing a smooth function $f(x^1,x^2)$ subject to $g(x^1,x^2) = 0$, where g is also smooth and where f and g are real valued. Suppose $\hat{x}$ provides a local solution to this problem. If $(\hat{g}_{x^1},\hat{g}_{x^2}) \neq 0$ then we can apply the implicit function theorem to solve for, say, x^2 uniquely in terms of x^1. Thus we have $g(x^1,\xi(x^1)) \equiv 0$ in a neighborhood of $\hat{x}^1$. So the constraint is always satisfied in that neighborhood. Our problem now is to maximize $\varphi(x^1) = f(x^1,\xi(x^1))$ locally and with no constraints in a sense to be made precise presently. By the 1st order necessary condition of Chapter 1 we have: $\hat{f}_1 + \hat{f}_2\hat{\xi}' = 0$. But $g(x^1,\xi(x^1))$ is a constant function around $\hat{x}^1$. Thus $\hat{g}_1 + \hat{g}_2\hat{\xi}' = 0$. Solving for $\hat{\xi}'$ we get: $\hat{\xi}' = - \hat{g}_1/\hat{g}_2$. Substituting, we have:

$$\hat{f}_1 = (\hat{f}_2)\ \hat{g}_1/\hat{g}_2$$

Assuming $\hat{f}_x \neq 0$ we get:

$$\hat{f}_1/\hat{f}_2 = \hat{g}_1/\hat{g}_2$$

Which means that $(\hat{f}_1\hat{f}_2)$ is proportional to $(\hat{g}_1\ \hat{g}_2)$, i.e. $\hat{f}_x = \lambda\ \hat{g}_x$ for some λ. We have backed into the "usual" way of presenting equality constrained maximization, except we have perhaps been a little more careful about stating conditions under which the "stuff" "works".

For instance suppose $\hat{g}_x = 0$, then our proof can't be carried out since we can't solve for either variable in terms of the others. The truth of the matter is even more dismal. The, yet to be formally stated theorem, is false if we do not assume that $\hat{g}_x \neq 0$. Consider the problem of maximizing $f(x) = x^1$ subject to $(x^1)^2 + (x^2)^2 = 0$. Clearly the maximum occurs at $(\hat{x}^1,\hat{x}^2) = (0,0)$. If the theorem was right then we would have: $1 = \hat{f}_{x^1} = \lambda \hat{g}_{x^1} = \lambda \cdot (0) = 0$ which is not very pleasant. (This example is due to Bliss (1946)).

We now state out problem more formally. Let $f: R^n \to R$ and $g^\alpha: R^n \to R$, $\alpha = 1, ..., m < n$, be C^2 functions. We wish to maximize f subject to $g(x) = 0$ where g is a vector function whose components are the g^α's. We say that $\hat{x}$ provides a *local solution to the equality constrained problem* (ECP) if there exists a neighborhood N of $\hat{x}$ such that $\hat{x} = \text{argu max } f$ on $N \cap \{x | g(x) = 0\}$. The set $\{x | g(x) = 0\}$ will be denoted by C_E and is referred to as the (equality) constraint set. When N is all of R^n we say that $\hat{x}$ is a *global* solution of ECP. The graphs below illustrate these definitions, $f(x) = a_i$ are level curves of f and where the arrow indicates higher level curves (higher a_i's).

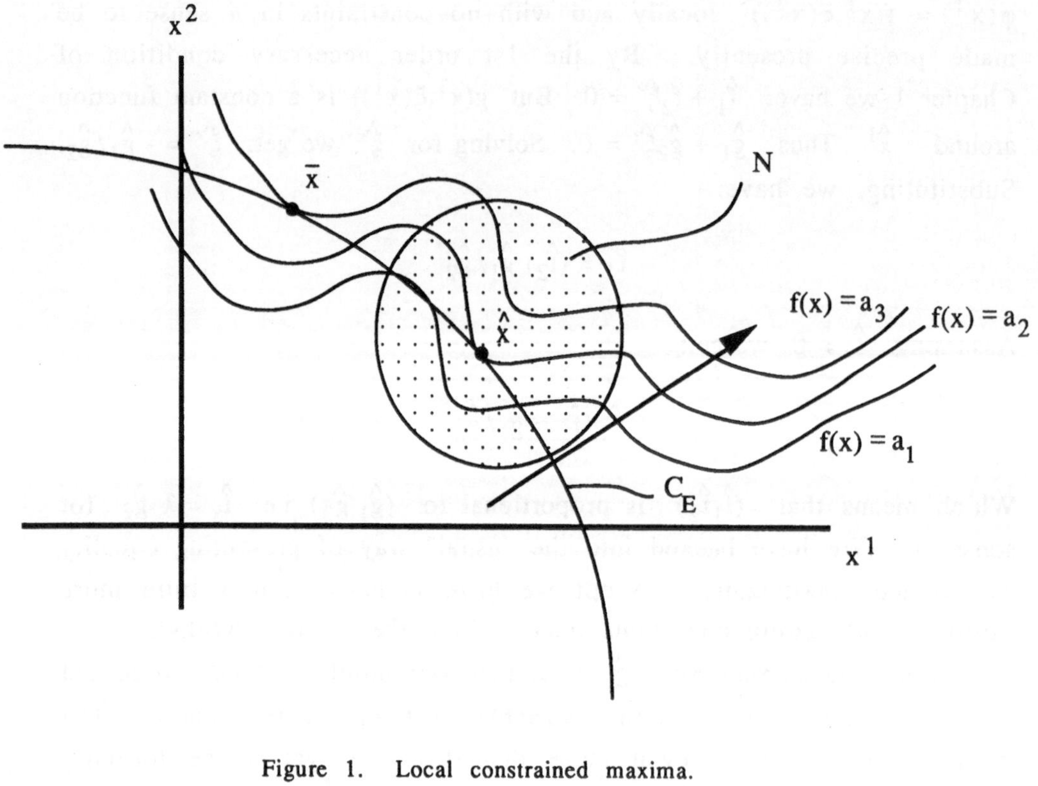

Figure 1. Local constrained maxima.

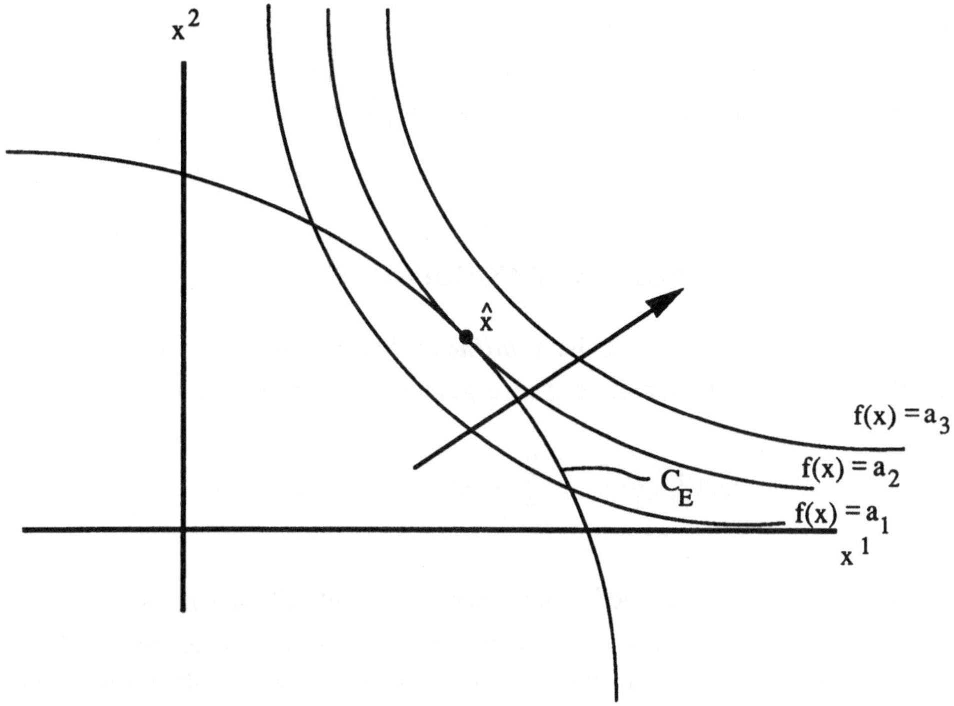

Figure 2. Global contained maximum.

In figure 1, $\hat{x}$ is a local solution of ECP but not a global one since $\bar{x}$ is on a higher level curve. In figure 2, $\hat{x}$ is a global solution to ECP.

A final introductory item. In our analysis in this and later chapters we will make extensive use of the implicit function theorem so it may be helpful to state it here. A proof can be found in any respectable real analysis book, e.g. Fleming (1977) or McShane and Botts (1954). Let $F: R^{n_1+n_2} \to R^{n_3}$, so that F is an n_3-vector valued function of an $n_1 + n_2$ vector say (u,v) where $u \in R^{n_1}$ and $v \in R^{n_2}$.

An implicit function theorem: Consider the function F *defined above. Assume that:* 1. $n_1 = n_3$. 2. *For some point* $(\mathring{u},\mathring{v})$, $F(\mathring{u},\mathring{v}) = 0$. 3. F *is* C^r *in* u *and* C^s *in* v, $r \geq 1$, $s \geq 0$ *in an open set containing* $(\mathring{u},\mathring{v})$. 4. *The Jacobian of* F *with respect* u *has a nonzero determinant at* $(\mathring{u},\mathring{v})$ i.e. $|\mathring{F}_u| \neq 0$. *Then there exist a constant* $\varepsilon > 0$, *a neighborhood* $N(\mathring{v})$, *and a unique function* μ *defined there such that:* i) $\mathring{u} = \mu(\mathring{v})$. ii) $F(\mu(v),v) = 0$ *for all* $v \in N(\mathring{v})$. iii) $\|\mu(v) - \mathring{u}\| < \varepsilon$ *for all* $u \in N(\mathring{v})$. iv) *The function* μ *is* C^s *on* $N(\mathring{v})$.

1. FIRST ORDER NECESSARY CONDITIONS

Let $\psi: R^{2(m+1)} \rightarrow R$ be a *bilinear functional*, i.e. a real valued function which is linear in either argument vector, given by:

$$\psi(\lambda^0,\lambda;y^0,y) = \sum_{\alpha=0}^{m} \lambda^\alpha y^\alpha = (\lambda^0,\lambda) \cdot (y^0,y),$$

where λ and y are in R^m and where λ^0 and y^0 are in R.

We discuss first, a characterization of solutions to our ECP in terms of the ψ which, when $(y^0,y) = (f,g)$, we shall term *pre-Lagrangian*. The first theorem we present, see Bliss (1938) or Bliss (1946), states the first order necessary condition in terms of the pre-Lagrangian.

Theorem 1. *If* $\hat{x}$ *is a local solution of* ECP *and if* f *and* g *are* C^1 *then there exists a* $(\lambda^0,\lambda) \neq 0$ *with* $\lambda^0 \geq 0$ *such that*

$$\hat{\psi}_\lambda = 0, \textit{ where } \hat{\psi}_\lambda = \hat{y} = \hat{g}, \textit{ and}$$

$$\hat{\psi}_x = 0, \textit{ where } \hat{\psi}_x = (\lambda^0,\lambda)\begin{pmatrix}\hat{f}_x\\ \hat{g}_x\end{pmatrix}.$$

The theorem applies to the Bliss example given in the introduction, where $f = x^1$ and $g = (x^1)^2 + (x^2)^2$. The theorem asserts the existence of a value $\hat{\lambda} = (\hat{\lambda}^0,\hat{\lambda}^1) \neq 0$ such that:

$$\hat{\lambda}^0\hat{f}_{x^1} = \hat{\lambda}^0 = \hat{\lambda}^1\hat{g}_{x^1} = 2\hat{\lambda}^1\hat{x}^1 = 0 \text{ and}$$
$$\hat{\lambda}^0\hat{f}_{x^2} = 0 = \hat{\lambda}^1\hat{g}_{x^2} = 2\hat{\lambda}^1\hat{x}^2 = 0.$$

The above equations hold e.g. for $\hat{\lambda} = (0,1)$.

So we try to prove the theorem since it does not look false. The theorem asserts that the system of linear equations:

$$(\lambda^0,\lambda)\begin{pmatrix}\hat{f}_x\\ \hat{g}_x\end{pmatrix} = 0$$

has a non-trivial solution: $(\hat{\lambda}^1,\hat{\lambda})$. Since $m < n$, by our maintained hypothesis, $m + 1 \leq n$ and the number of unknowns does not exceed the number of equations. Using a theorem from linear algebra, all we have to show is that the rank of $\begin{pmatrix}\hat{f}_x\\ \hat{g}_x\end{pmatrix}$ is less than $m + 1$. We will do that now. Assume, by way of contradiction, that the rank of that matrix is not less than $m + 1$. Then it has to be equal to $m + 1$ since it can't be larger (again because $m + 1 \leq n$). We can then, by using the implicit function theorem show that we could locally extend the solution of the constraint equations around $\hat{x}$ and get a higher value for f than at $\hat{x}$ and thus contradict the assertion that $\hat{x}$ solves the ECP. To this end, renumber the components of x so that the first $m + 1$ columns of $\begin{pmatrix}\hat{f}_x\\ \hat{g}_x\end{pmatrix}$ are linearly independent. We know that exactly $m + 1$ columns of our matrix are linearly independent since its rank is $m + 1$. Denote the vector whose components are the first $m + 1$ components of the renumbered x components by x^I. Denote the vector with the remaining components of x by x^{II}. Let the function $h: R^{n+1} \to R$ be given by:

$$h(x,t) = f(x) - f(\hat{x}) - t.$$

we shall apply the implicit function theorem stated above, to the function $\begin{pmatrix}h\\ g\end{pmatrix}$ with $u = x^I$ and $v = (x^{II},t)$ with $(\mathring{u},\mathring{v}) = (\hat{x},0)$. If the implicit function theorem applies then we would get a neighborhood of $\hat{x}$ in R^n and a neighborhood M of O in R such that $g(x) = 0$ and $f(x) = f(\hat{x}) + t$ for all $x \in$

M and for all t in an interval about zero. Taking $(\bar{x}, \bar{t})$ so that $\bar{x} \in N \cap M$ and $\bar{t} > 0$ we have found a point $\bar{x}$ in the comparison set $N \cap C$ of $\hat{x}$ with $f(\bar{x}) > f(\hat{x})$ which is the sought contradiction. But the implicit function theorem does apply. First $\begin{pmatrix} h(\hat{x},0) \\ g(\hat{x}) \end{pmatrix} = 0$ by definition of h and since $\hat{x}$ satisfies the constraints. The functions h and g are smooth enough by virtue of our maintained hypothesis that f and g are C^2 and by definition of h. Finally, the Jacobian $\begin{pmatrix} \hat{h}_{xI} \\ \hat{g}_{xI} \end{pmatrix}$ has rank m + 1 by the contradiction hypothesis. □

If we apply theorem 1 to a constrained maximization problem then we could obtain a characterization of the solution which is independent of the maximand, which happens when $\hat{\psi}_x = 0$ holds with $\lambda^0 = 0$. This can occur when the constraint set is a single point as in the Bliss example, presented above. Another possible reason for such an embarrassing occurrence is that some constraints may be redundant. At any rate, it is useful to provide sufficient conditions for λ^0 to be nonzero. Such conditions can be problem specific, in that the statement of the economic problem should contain enough information to prove directly that $\lambda^0 \neq 0$. On the other hand they can be stated as a general condition that the constraints should satisfy. The case where theorem 1 holds with $\lambda_0 > 0$ was termed *normal* by Bliss, see Bliss (1938), and we will state and prove Bliss's modification of theorem 1 now.

Theorem 2. *If in addition to the assumptions of theorem 1 we have: the rank of the matrix* $\hat{g}_x$ *is equal to* m. *Then there exists a unique vector* $\lambda \in R^m$ *such that* $\hat{F}_x = 0$, *where* $F = f + \lambda g$.

By theorem 1 there exists a nonzero vector $(\tilde{\lambda}^0,\tilde{\lambda})$ such that $(\tilde{\lambda}^0,\tilde{\lambda})\begin{pmatrix}\hat{f}_x \\ \hat{g}_x\end{pmatrix} = 0$. If $\tilde{\lambda}^0$ is zero then $\lambda\hat{g}_x = 0$ which is a system of n equations in m unknowns has a nontrivial solution $\tilde{\lambda}$. But this implies that the rank of $\hat{g}_x$ is less than m, since $m < n$. Thus $\tilde{\lambda}^0 > 0$. Let $\lambda = (1/\tilde{\lambda}^0)\,\tilde{\lambda}$. Then $\hat{F}_x = 0$. To prove uniqueness, suppose $\bar{\lambda} \neq \lambda$ is such that $\hat{\bar{F}}_x = \hat{f}_x + \bar{\lambda}\,\hat{g}_x = 0$. Then $\hat{F}_x - \hat{\bar{F}}_x = (\lambda - \bar{\lambda})\hat{g}_x = 0$. Thus the equation system $z\hat{g}_x = 0$ has a non trivial solution $(\lambda - \bar{\lambda})$. But this, again, is a contradiction since the rank of $\hat{g}_x$ is m.□

<u>Example 1:</u> (Bliss (1938)). Let $f(x^1,x^2)$ have nonzero derivatives at $x = 0$. Let $g(x) = (x^1)^2 + (x^2)^2$, then $C = \{0\}$ and $\hat{x} = 0 =$ argu max f on C. By theorem 1 there exists $(\lambda^0,\lambda) \neq 0$ such that $\hat{\lambda}^0\hat{f}_{x^1} + 2\hat{\lambda}^1\hat{x}^1 = \hat{\lambda}^0\hat{f}_{x^1} = 0$ and $\hat{\lambda}^0\hat{f}_{x^2} + 2\hat{\lambda}^1\hat{x}^2 = \hat{\lambda}^0\hat{f}_{x^2} = 0$, so $\hat{\lambda}^0\hat{f}_x = 0$ thus $\hat{\lambda}^0 = 0$ since $\hat{f}_x \neq 0$. We may take λ to be 45.67 since it is arbitrary. Theorem 2 does not apply since $\hat{g}_x = (0,0)$ does not have rank 1 as required by that theorem.

<u>Example 2:</u> (Bliss (1938)). Let $(x^1,x^2) = -(x^2)^2 - 2(x^1)^2$, and let $g(x^1,x^2) = (x^1)^2x^2 - (x^2)^3$. The constraint set is given by:

$$C = \{(x^1,x^2) \mid x^1 = x^2 \text{ and } x^2 \neq 0\} \cup \{(x^1,x^2) \mid x^2 = 0\}$$

and $\hat{x} = 0 =$ argu max f on C. Theorem 1 applies and we have: there exists $(\lambda^0,\lambda) \neq 0$ such that $-4\lambda^0\hat{x}^1 + 2\lambda\hat{x}^1\hat{x}^2 = 0$ and $2\lambda^0\hat{x}^2 + \lambda((\hat{x}^1)^2 - 3(\hat{x}^2)^2) = 0$. The equations do not impose any conditions on (λ^0,λ) since $\hat{x} = 0$ so we take $(\lambda^0,\lambda) = (1,9.3)$. In fact we verify the conclusion of theorem 1 and theorem 2 even though theorem 2 does not apply since $\hat{g}_x = (2\hat{x}^1\hat{x}^2 \quad (\hat{x}^1)^2 - 2(\hat{x}^2)^2) = (0\ 0)$ and thus has zero rank.

Exercise 1. Verify theorem 2 for $f(x) = x^1x^2x^3$ and $g(x) = 3 - x^1 - x^2 - x^3$, $\hat{x} = (1,1,1)$.
Exercise 2. Verify theorem 2 for f and g as in Exercise 1 and added constraint $x^1 = x^2$, $\hat{x} = (1,1,1)$. Same λ?
Exercise 3. Verify the first order necessary conditions (all of them) for $f = 6 - 4x^1 - 3x^2$ and $g = 1 - (x^1)^2 - (x^2)^2$.

2. UTILITY MAXIMIZING CONSUMERS 2

We assume that a household makes decisions about consumption by maximizing a single utility function $u(x)$, where $x \in R^n_+$ is a commodity bundle, subject to the budget constraint given by: $m - px = 0$, where $m \in R_+$ is the amount of liquidity available to finance consumption purchases and where $p \in R^n_+$ is a price vector taken parametrically by the consumer. The subscript $+$ denotes non-negative elements of the respective space. The number m, assumed positive here, is confusingly referred to as income by economists, and in compliance with tradition we will preserve the confusion and call m income but will come back with comments later.

Assume that the utility function is smooth, that $p \neq 0$, *and that* $\hat{x} > 0$ *solves the consumer's decision problem. Then there exists a unique* λ *such that*
$\hat{u}_x = \lambda p$.

The above assertion will be proved in two ways. First we prove it using theorem 1. The theorem applies, so there exist $(\lambda^0, \bar{\lambda}) \neq 0$ such that $\lambda^0 \hat{u}_x = {}^{\pm}\bar{\lambda} p$. We note that $\lambda^0 \neq 0$, for if it were then ${}^{\pm}\bar{\lambda} p = 0$

where $^{\pm}\bar{\lambda} \neq 0$ which implies that $p = 0$ contradiction the assumption that $p \neq 0$. Thus we take $\lambda = {}^{\pm}\bar{\lambda}/\lambda^0$. If there was another λ, say λ' which satisfies $\hat{u}_x = \lambda p$ then we would have $(\lambda - \lambda')p = 0$. Since $p \neq 0$ we must have $\lambda = \lambda'$ so λ' could not have been different from λ.

The other way of proving the assertion is to apply theorem 2 which applies since $g(x) = m - px$ and $g_x = -p$ which has full rank since $p \neq 0$.

3. FIRST ORDER SUFFICIENT CONDITIONS

If the maximand is concave and the constraints are affine then a critical point of the pre-Lagrangian is a point which provides a global solution of the ECP. We can't do better as far as the constraints are concerned, i.e. concavity of g won't do. The reason is that the constraint set is unchanged if we replace g by - g (that is why the sign of λ in theorem 2 is left unspecified) and then our constraints turn convex on us, but that won't matter if they are affine. Formally:

Theorem 3. *Let* f *be smooth and concave and let* g *be affine. If* $\hat{x}$ *is such that* $\hat{\psi}_x = 0$, *with* $\lambda^0 > 0$, *where* $\psi = (\lambda^0, \lambda)\binom{f}{g}$, *and if* $\hat{\psi}_\lambda = 0$ *then* $\hat{x}$ *is a global solution of* ECP.

To prove the above theorem note that ψ is concave and smooth. By the first order sufficient condition of Chapter 1, ψ is maximized at $\hat{x}$. Thus

$$\lambda^0 f(\hat{x}) + \lambda g(\hat{x}) \geq \lambda^0 f(x) + \lambda g(x) \qquad \forall\, x.$$

now let x satisfy the constraints. Then $g(x) = 0$. Since $g(\hat{x}) = 0$ we have

$$\lambda^0 f(\hat{x}) \geq \lambda^0 f(x)$$

But then $f(\hat{x}) \geq f(x)$.□

4. SECOND ORDER NECESSARY CONDITIONS

Before we state the second order necessary conditions, which play an important role in comparative statics and all manner of sensitivity analysis that economists undertake, we have to make some introductory remarks. The first order condition characterizes the change in the function near the extreme point so that in the unconstrained case the velocity where movement is admissible in any direction. In the constrained case, motion must occur on a "surface" determined by the constraints. They characterize derivatives of the maximand on a manifold. Second order conditions characterize changes in "velocity" of motion so they characterize "acceleration". In the constrained case, to continue with our metaphor, we talk of acceleration where motion is constrained. Thus we talk of second differentials of the maximand on the constraint manifold, and so our increments are constrained to be in the tangent space, at solution point, to the constraint manifold. Geometrically, as if the above was not geometric enough, the graph below may explain the idea:

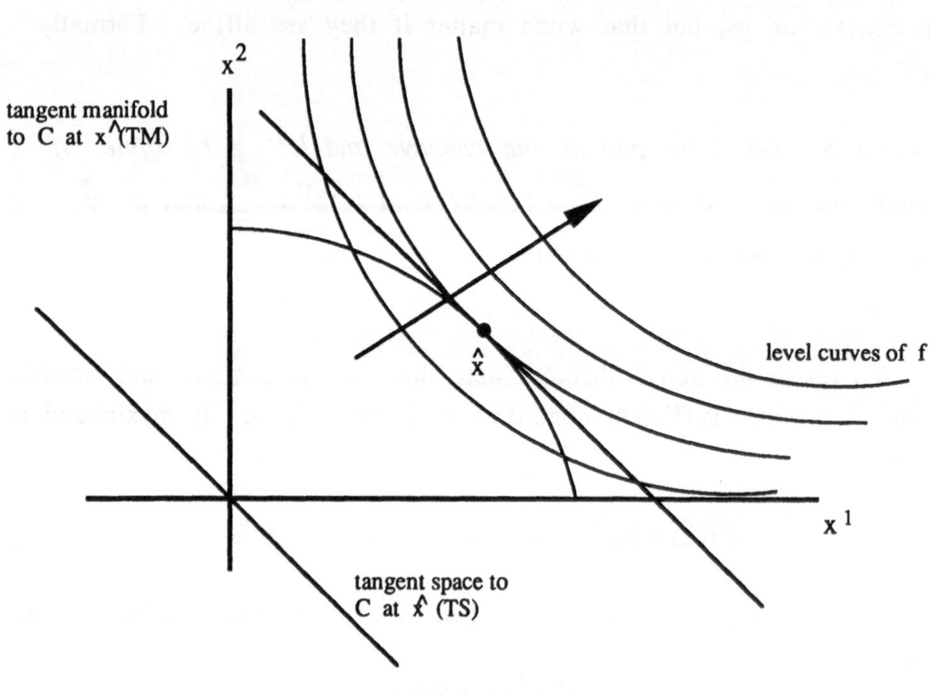

Figure 3. Tangent manifold and tangent space to C at $\hat{x}$.

In the section we show that a necessary condition for $\hat{x}$ to solve ECP is that the second differential of the Lagrangian function, $\eta \hat{F}_{xx} \eta^*$, be non-positive for η satisfying $\hat{g}_x \eta^* = 0$. The method of proof is to parametrize the constraint set, i.e. express the point x that satisfies $g(x) = 0$ as function of a real variable, and then apply the second order necessary condition for a nonconstrained maximum of a function of one real variable. We first state theorem 4, then a parametrization lemma which we use in the proof of theorem 4. Then we prove the lemma.

Theorem 4. *If* 1) f *and* g *are of class* C^2. 2) $\hat{x}$ *is a local solution to* ECP. 3) *the rank of* $[\hat{g}_x]$ *is* m. *Then* $\eta \hat{F}_{xx} \eta^* \leq 0$ for all η with $\hat{g}_x \eta^* = 0$, *where* "*" *denotes the transpose.*

Lemma: *(parametrizability). If conditions* 1) *and* 3) *of theorem 4 hold and if* $\eta = x - \hat{x}$ *satisfies* $\hat{g}_x \eta^* = 0$ *then there exists an n-vector valued function* $\omega(b)$ *of a real variable* b *such that:*

(1) $\omega(b)$ *is of class* C^2 *in a neighborhood* S *of* $b = 0$.

(2) $g[\omega(b)] = 0$ *for* $b \in S$, *i.e. if we substitute* $x = \omega(b)$ *in* $g(x)$ *then the constraints are satisfies for all* $b \in S$.

(3.a) $\omega(0) = \hat{x}$.

(3.b) $\omega'(0) = \eta$, *where* "′" *denotes the derivative.*

Proof of theorem 4: Let η satisfy $\hat{g}_x \eta^* = 0$. We wish to show that $\eta \hat{F}_{xx} \eta^* \geq 0$. By the lemma, we can write $x = \omega(b)$ for $b \in S$ where $\omega(0) = \hat{x}$ and $g[\omega(b)] = 0$. Thus, $b = 0$, by hypothesis of the theorem, provides a local unconstrained maximum for $\phi(b) = f[\omega(b)]$ in S. Hence:

(i) $\phi'(0) = 0$.

(ii) $\phi''(0) \leq 0$.

But, using the chain rule,

(iii) $\phi'(b) = \sum_{i=1}^{n} f_i[\omega(b)]w_i'(b)$.

Now, consider Lagrangian function F whose derivatives with respect to x are the components of the vector:

(iv-1) $F_x = f_x + \lambda g_x$

Post-multiplying both sides of (iv-1) by $(w'(b))^*$, i.e. by the column vector whose components are $w_i'(b)$, and substituting $x = \omega(b)$ we have:

(iv-2) $F_x(\omega')^* = f_x(\omega')^* + \lambda g_x(\omega')^*$.

By lemma 1, $g[\omega(b)] = 0$ for $b \in S$. Thus $\frac{d}{db} g[\omega(b)] = g_x(\omega'(b))^* = 0$ for $b \in S$. Premultiplying by λ we have:

(v) $\lambda g_x(\omega'(b))^* = 0, \qquad b \in S$.

By (iv-2) and (v) we have: $F_x(\omega') = f_x(\omega')^*$, thus, by (iii), $\phi'(b) = F_x(\omega'(b))^*$. Differentiating the last expression with respect to b we get: $\phi''(b) = \omega'(b)F_{xx}(\omega'(b))^* = F_x(\omega''(b))^*$, where $(\omega''(b))^*$ is the n-column vector whose components are the second derivatives $\omega''(b)$.

Evaluation at $b = 0$, we get:

(vi) $\phi''(0) = \omega'(0)F_{xx}[\omega(0),\lambda](\omega'(0))^* + F_x[\omega(0),\lambda](\omega''(0))^*$.

But $F_x[\omega(0),\lambda] = \hat{F}_x = 0$, by first order necessary condition, thus the second term of the right hand side of (vi) is zero. Also, by lemma, $\omega'(0) = \eta$ and $\omega(0) = x$. Hence we may write (vi) as:

(vii) $\phi''(0) = \eta\hat{F}_{xx}\eta^*$.

By (ii) and (vii) the proof of the theorem is complete, since η is an arbitrary vector with $\hat{g}_x\eta^* = 0$.□

Proof of the lemma: Let $\eta = x - \hat{x}$ satisfy $\hat{g}_x \eta^* = 0$. By hypothesis of the lemma, the matrix g_x has rank m. Without loss of generality we may assume that the columns $\hat{g}_{x_s}$ are linearly independent, s = 1, ..., m. Let the matrix $\hat{g}_{x^I}$ denote the matrix formed by these columns. Denote by $\therefore o(g,\hat{\ })_{x_\gamma}$, the remaining columns of $\hat{g}_x$ and by $\hat{g}_{x^{II}}$ the matrix formed by them. The vector x is now partitioned into (x^I, x^{II}), similarly we partition η into $\eta = (\eta^I, \eta^{II})$. The assertion of the lemma is that we can express x as a function of b. Consider the n-vector function $G(x,b) = G^I(x,b), G^{II}(x,b))$, with G^I having m components and G^{II} having n - m components, where $G^I(x,b) = g(x)$ and $G^{II}(x,b) = x^{II} - (b\eta^{II} + \hat{x}^{II})$. We shall apply the implicit function theorem to $G(x,b)$ with "initial point" $(x,b) = (\hat{x},0)$. Note that $G(\hat{x},0) = (g(\hat{x}), \hat{x}^{II} - \hat{x}^{II}) = 0$. Note that the Jacobian $\hat{G}_x$ is: $\begin{bmatrix} \hat{g}_{x^I} & \hat{g}_{x^{II}} \\ 0 & I \end{bmatrix}$ where 0 is an $n - m \times m$ zero matrix and I is $n - m \times n - m$ identity matrix. Clearly $\hat{G}_x$ has rank n. The functions $G(x,b)$ are of class C^2 by hypothesis and by construction. Thus the implicit function theorem applies and we can obtain $x = \omega(b)$ for b in some neighborhood of $b = 0$. By conclusions 1), 2) and (3.a) of our lemma. It remains to prove (3.b). In $G^I(x,b)$ set $x = \omega(b)$. Thus $G^I(\omega(b),b) = g[\omega(b)] = 0$, $b \in S(0)$. differentiating with respect to b we have:

(i) $\frac{d}{db} g[\omega(b)] = g_{x^I}(\omega'^I)^* + g_{x^{II}}(\omega'^{II})^* = 0.$

Now $\omega^{II}(b) = \hat{x}^{II} + b\eta^{II}$, by solving for x^{II} from $G^{II}(x, b) = 0$. Thus:

(ii) $\omega'^{II}(b) = \eta^{II}$.

By (ii) we have: $\omega^{II}(0) = \eta^{II}$, so it remains to show that $\omega'^I(0) = \eta^I$. Substitute from (ii) into (i) getting: $\frac{d}{db} g[\omega(b)] = g_{x^I}(\eta^I)^* + g_{x^{II}}(\eta^{II}))^* = 0$. Evaluating at $b = 0$, we get

(iii) $\hat{g}_{x^I}(\omega'^I(0))^* + \hat{g}_{x^{II}}(\omega'^{II}(0))^* = 0.$

Since η satisfies $\hat{g}_x\eta^* = 0$, we have:

(iv) $\hat{g}_x\eta = \hat{g}_{x\mathrm{I}}(\eta^{\mathrm{I}})^* + \hat{g}_{x\mathrm{II}}(\eta^{\mathrm{II}}))^* = 0.$

Subtracting (iv) from (iii) we have:

(v) $\hat{g}_{x\mathrm{I}}(\omega'^{\mathrm{I}}(0) - \eta^{\mathrm{I}})^* = 0.$

But (v) is a system of linear homogeneous equations with a square nonsingular matrix of coefficients. Hence (v) can only have the trivial solution $\omega'^{\mathrm{I}}(0) - \eta^{\mathrm{I}} = 0$, i.e. $\omega'^{\mathrm{I}}(0) = \eta^{\mathrm{I}}$. Thus $\omega'(0) = \eta$ as we were to prove.□

5. SECOND ORDER SUFFICIENT CONDITIONS

In this section we prove that if, in addition to the vanishing of its first derivatives, the second differential of the pre-Lagrangian ψ is positive definite for $\eta \neq 0$ with $\hat{g}_x\eta^* = 0$ then the point $\hat{x}$ is a local solution of ECP.

Theorem 5. *If* 1) f *and* g *are of class* C^2. 2) $\therefore o(\psi, \hat{\ })_x = 0$ *at some point* $\therefore o(x, \hat{\ })$ *with* $g(\therefore o(x, \hat{\ })) = 0$. 3) $\eta \therefore o(\psi, \hat{\ })_{xx} < 0$ *for* η *satisfying* $\hat{g}_x\eta^* = 0$. *Then* $\hat{x}$ *is a local solution of* ECP.

Proof: The theorem is proved by contradiction. Suppose $\hat{x}$ is not a local solution to ECP. Then there exists a sequence of distinct points $\{x_r\}$ with $x_r \neq \hat{x}$ converging to $\hat{x}$ such that $g(x_r) = 0$ and $f(x_r) \geq f(\hat{x})$. Define $k_r = \|x_r - \hat{x}\|$. Then $k_r \to 0$, since $x_r \to \hat{x}$. Define $h_r^i = \dfrac{x_r^i - \hat{x}^i}{k_r}$, $i = 1, \ldots, n$, and let $h_r = (h_r^1, h_r^2, \ldots, h_r^n)$. Note that $\|h_r\| = \dfrac{1}{k_r}\|x_r\| = 1$. The sequences $\{h_r^i\}$ are bounded, hence each one of them has a convergent subsequence. We shall abuse the notation and refer to these convergent subsequences as $\{h_r^i\}$. Denote the limits of $\{h_r^i\}$ by h_0^i. Our contradiction consists of showing that $h_0^i = 0$, i.e. we have a sequence h_r with $||h_r|| = 1$ that converges to the zero vector.

Now $g(x_r) = 0$, thus $\psi(x_r) = \lambda_0 f(x_r)$. By the contradiction assumption $f(x_r) - f(\hat{x}) \geq 0$. Since $\lambda_0 \geq 0$, we have $\lambda_0 f(x_r) - \lambda_0 f(\hat{x}) \geq 0$. But $\psi(\hat{x}) = \lambda_0 f(\hat{x})$. Thus:

(1) $\psi(x_r) - \psi(\hat{x}) \geq 0.$

The elements of $\{x_r\}$ are distinct from $\hat{x}$, hence $k_r > 0$. Thus:

(2) $\frac{1}{k_r^2}(\psi(x_r) - \psi(\hat{x}) \geq 0.$

By Taylor's theorem:

(3) $\psi(x_r) - \psi(\hat{x}) = \hat{\psi}_x \eta_r^* + \frac{1}{2}\hat{\psi}_{xx}\eta_r^* + R.$

where R is of smaller order of magnitude than $||x_r - \hat{x}||^2$ as $||x_r - \hat{x}|| \to 0$. By hypothesis $\hat{\psi}_x = 0$, hence:

(4) $\psi(x_r) - \psi(\hat{x}) = \frac{1}{2}\eta_r\hat{\psi}_{xx}\eta_r^* + R.$

Dividing both sides of (4) by k_r^2 and noting that $h_r = \frac{\eta_r}{k_r}$ we have:

(5) $\frac{1}{k_r^2}(\psi(x_r) - \psi(\hat{x}) = \frac{1}{2} h_r\hat{\psi}_{xx}h_r^* + \frac{R}{k_r^2}.$

By (2) and (5) we have:

(6) $h_r\hat{\psi}_{xx}h_r^* + 2\frac{R}{k_r^2} \geq 0.$

Passing to the limit through the sequences $\{h_r\}$ we have:

(7) $h_0\hat{\psi}_{xx}h_0^* \geq 0$.

We now show that $h_0 = 0$. This will be done by showing that h_0 satisfies $\hat{g}_x h_0^* = 0$, for in that case either $h_0 = 0$ or $h_0\hat{\psi}_{xx}h_0^* < 0$ with the later possibility contradicting (7). Since $g(x_r) = g(\hat{x}) = 0$, we have $\frac{1}{k_r}[g(x_r) - g(\hat{x})] = 0$. By the mean value theorem we have:

(8) $0 = \frac{1}{k_r}[g^{\alpha}(x_r) - g^{\alpha}(\hat{x})] = \frac{1}{k_r} g_x^{\alpha}(\hat{x} + \theta\eta_r)\eta_r^* = g_x^{\alpha}(\hat{x} + \theta\eta_r)h_r^*$.

By continuity of g_x^{α}, taking the limit as the sequences $\{h_r\} \to \{h_0\}$ we have, since $\hat{x} + \theta\eta_r \to \hat{x}$,

(9) $\hat{g}_x h_0^* = 0$.

By (9) the proof is complete.□

6. UTILITY MAXIMIZING CONSUMERS 3

Recall our consumer of section 4. We define the *demand map* of that consumer as: $h(p,m)$ = argu max $u(x)$ on $B(p,m)$, where $B(p,m) = \{x \in R^n | \ m - px\} = 0$.

We maintain the hypothesis that u is smooth and assume in addition that $p > 0$ and $m > 0$. We can, and will, show that $h(p,m)$ is a differentiable function which satisfies the usual conditions about demand functions.

<u>Proposition</u>: *If the utility function is* C^2 *and if it has a nonzero second differential,* i.e. $\xi u_{xx}\xi^* \neq 0$ *for* $\xi \neq 0$, *then the demand map is a continuously differentiable function and satisfies the following conditions:*

1. *It is homogeneous of degree zero in prices and liquidity,* i.e. *in* p *and* m.
2. *It satisfies the budget equation.*

3. *The Slutskii matrix is negative semi-definite.*
4. *The Slutskii matrix is symmetric.*

To Prove the proposition we apply the implicit function theorem to:

(1) $$\begin{pmatrix} u_x - \lambda p \\ m - px \end{pmatrix} = R(x,p,m)$$

where $R(x,p,m) = 0$ whenever x solves the consumer maximization problem (see section 4). We use the implicit function theorem to solve (1) for x and λ in terms of p and m. The theorem applies since $\begin{pmatrix} u_{xx} & -p^* \\ -p & 0 \end{pmatrix}$, the Jacobian of (1) is nonsingular by assumption. Thus there exist continuously differentiable functions $h(p,m)$, $\ell(p,m)$ defined in a neighborhood of wherever we started such that

(2.1) $$u_x(h(p,m) - \ell(p,m)p) = 0$$

(2.2) $$m - p\, h(p,m) = 0$$

Since the equality in the second order necessary condition is excluded, by assumption, (2) implies that $h(p,m)$ locally solves the consumer maximization problem because the second order necessary condition is satisfied. Thus $h(p,m)$ is a demand map. That it is a function and a differentiable one at that, follows from the implicit function theorem. Condition (2.2) implies that h satisfies the budget equation.

The terms of the Slutskii matrix are defined to be the compensated changes in the demand for good i when the price of good j is changed. Following Slutskii himself we take "compensated" to mean that the income of the consumer is adjusted to a level that enables him to afford the bundle which he bought before the price changed. Denote the compensated income by $\mu(p)$ and the compensated demand function by $H(p)$ so that $H(p) = h(p,\mu(p))$. Denote the Slutskii matrix by $K = (k_{ij})$,

$$K_{ij} = \frac{\partial H^i}{\partial P_j} = \frac{\partial h^i}{\partial p_j} + \frac{\partial h^i}{\partial m} + \frac{\partial \mu}{\partial p_j}.$$

To compute $\frac{\partial \mu}{\partial p_j}$ note that $p \cdot dx = 0$, by compensation, so:

$$d\mu = d(p \cdot x) = p \cdot dx + x \cdot dp = x \cdot dp$$

Thus $\frac{\partial \mu}{\partial p_j} = x^j$. We have:

(3) $$K_{ij} = h^i_j + h^i_m x^j.$$

To prove (3) we must show that:

$$\sum_{i,j} K_{ij}\, a_i\, a_j \leq 0 \quad \text{for any vector } a.$$

Let a be given and take $dp = a$. Taking the differential of H_i we get:

$$dH^i = \sum_i (h^i_j + h^i_m) dp_j = \sum_j K_{ij} dp_j$$

Thus

$$\sum_i dH^i dp_i = \sum_i \sum_j K_{ij}\, dp_i\, dp_j$$

We prove 3. if we prove $\sum_i dH^i\, dp_i \leq 0$. Since (2.1) holds as an identity near the initial x, the differential of the left hand side is zero. So:

$$\sum_i u_{ij}\, dH^j - \ell\, dp_i + p_i\, dl = 0$$

Thus

(4) $$\sum_i \sum_j u_{ij}\, dH^i\, dH^j = \ell \sum dp_i\, dH^i + d\ell \sum p_i\, dH^i$$

But $\sum p_i dH^i = 0$ by compensation and $\ell \geq 0$ by (2.1). Thus $dH \cdot dp$ has the same sign as the left hand side of (4) which is negative semi-definite by the second order necessary condition since $p \cdot dH = 0$. To show that 4. is true use H instead of h in (2) and use the implicit function theorem to calculate $[H^i_j] = [K_{ij}]$. Then note that the formula involves the inverse of a symmetric matrix, since u is C^2, and conclude that $[K_{ij}]$ is symmetric.

An easier proof of 4. in the proposition is to note that $\mu(p)$, the compensated income function is C^2 and since, by Young's theorem (Young 1909) we have: $\mu_{ij} = \mu_{ji}$. But $K_{ij} = H^i_j = K_{ij}$ and the proof would be complete. Indeed, by (2.2), and the implicit function theorem:

$$\mu_i = H^i$$

Thus μ_i is C^1 hence μ is C^2.

7. COST FUNCTIONS

A firm uses n inputs, x, to produce a single output and does so "efficiently" in a sense to be made clear later and so its technological constraint is given by: $y = f(x)$, where f is the firm's production function. We define the firm's *cost function*, $\varphi(y)$, as follows:

$$\varphi(y) = \min p \cdot x \text{ subject to } y = f(x).$$

Assume that f is C^2 and let x solve the problem associated with the definition of our cost function. Assume further that $f_x \geq 0$, a reasonable input has positive marginal product. Then by theorem 2 there uniquely exist λ such that:

$$\begin{aligned} p &= - f_x \\ y &= f(x) \end{aligned} \qquad (1)$$

Assuming that $\begin{pmatrix} f_{xx} & f_x^* \\ f_x & 0 \end{pmatrix}$ is nonsingular we apply the implicit function theorem and conclude the existence of continuously differentiable functions $\xi(y,p)$, $\ell(y,p)$ such that

(2.1) $p = \ell(y,p)\ f_x\ (\xi(y,p)$

(2.2) $y = f(\xi(y,p)$

Assuming further that stationarity implies minimization (e.g. that the second differential of f is negative since our Lagrangian $F = - p \cdot x + \lambda(f(x) - y)$, then $\xi(y,p)$ is an outlay minimizing conditional demand function for inputs. The cost function then satisfies:

(3) $\varphi(y,p) = p \cdot \xi$

Note that φ is continuously differentiable and the marginal cost, $\frac{\partial \varphi}{\partial y}$, is given by:

(4) $\frac{\partial \varphi}{\partial y} = \ell(y,p)$

By (4), differentiating again,

(5) $\frac{\partial^2 C}{\partial y} = \frac{\partial \ell}{\partial y}$.

By taking total derivative of (2.) we have

$$0 = (dl)f_x + \ell\ df_x$$

Thus

(6) $dl\ dy = - \ell\ d^2 f$

By (2.1), if $f_x \geq 0, \ell \geq 0$, and so (6) implies that marginal cost decreases when f is convex and increases when f is concave.

8. FURTHER NOTES ON SECOND ORDER CONDITIONS

We characterize second order conditions in terms of bordered Hessians and related second order conditions to concavity concepts. First we state a characterization of definite quadratic forms under linear constraints (see e.g. Caratheodory (1935) or Shostak (1954)). Consider the quadratic form: $Q(A,y) = yAy^*$, where A is an $n \times n$ symmetric matrix and where $y \in R^n$. We wish to characterize the definiteness of Q subject to: $By = 0$, where B is an $m \times n$ matrix of rank m. let D be defined as follows:

$$D = (-1)^m \left[\begin{array}{c|c} 0 & B \\ \hline B^* & A \end{array}\right]$$

where 0 is a square zero matrix of order m.

A necessary and sufficient condition for the quadratic form Q *to be positive definite subject to* $By = 0$ *is than the principal diagonal minors of* D *of order* k *are positive, where* $k = 2m + 1, 2m + 2, \ldots, m + n$. *For negative definiteness the condition is that the principal minors of order* k *have the sign of* $(-1)^k$, $k = 2m + 1, \ldots, m + n$. *In order for* Q *to be positive semi-definite it is necessary and sufficient that each of the principal diagonal minors of* D *of order* k *is zero or positive for* $k = 2m + 1, \ldots, m + n$. *For negative semi-definite the condition is that the principal diagonal minors of order* k *either vanish or have the sign of* $(-1)^k$, $k = 2m + 1, \ldots, m + n$.

Example. Let $A = \begin{bmatrix} 0 & 1 \\ 1 & 0 \end{bmatrix}$ and let $B = [1 \quad 1]$. The quadratic form is:

$Q = [y^1 \quad y^2] \, A \begin{bmatrix} y^1 \\ y^2 \end{bmatrix} = 2y^1y^2$ which is *not* definite. But if constrained by $By = 0$ so that $y^1 + y^2 = 0$ then upon substitution we get: $Q = -2(y^1)^2$ which is negative definite.

Using our characterization we have:

$$D = (-1)\begin{bmatrix} 0 & 1 & 1 \\ 1 & 0 & 1 \\ 1 & 1 & 0 \end{bmatrix} = \begin{bmatrix} 0 & -1 & -1 \\ -1 & 0 & -1 \\ -1 & -1 & 0 \end{bmatrix}$$

The first (and last) order of principal diagonal minors that we must look at are of order 3 and there is only one of them. It has to either vanish or have the sign of $(-1)^3$ (so has to be negative). That determinant is $|D| = 0 - 1 - 1 + 0 + 0 + 0 = -2 < 0$ comme il faut.

We can use the above characterization of definiteness of quadratic forms to restate our second order conditions and we shall so do. The "coefficients" of the second differential of a continuously differentiable function $f\colon\ R^n \to R$ are given by the symmetric matrix $H = [F_{xx}]$ which is the Hessian of that function introduced earlier. If the differentiation occurs on the manifold, surface, given by $g(x) = 0$, where $g : R^n \to R^m$ is smooth, then the "coefficients" that are relevant are related to the matrix:

$$\mathbb{H}(\overset{\circ}{x}) = (-1)^m \begin{bmatrix} 0 & \overset{\circ}{g}_x \\ \overset{\circ}{g}{}^*_x & \overset{\circ}{F}_{xx} \end{bmatrix} = \overset{\circ}{\mathbb{H}},$$

which is called the bordered Hessian at $\overset{\circ}{x}$ by some economists and which should be called a constrained Hessian but since Hessions are known to be hard to constrain we let the suggestion die with dignity at birth.

Theorem 6: *Let* f, g *and* $\hat{x}$ *be as in theorem 4. Then all the principal diagonal minors of* $\hat{\mathbb{H}}$ *of order* k *are either zero or have the sign of* $(-1)^k$, *where* $k = 2m + 1, ..., m + n$, *and where* $F = f + \lambda g$.

Theorem 7. *Let* f, g *and* $\hat{x}$ *satisfy conditions* 1) *and* 2) *of theorem* 5. *Suppose in addition that the rank of* $\hat{g}_x$ is m. *The point* $\hat{x}$ *solves ECP locally if the principal diagonal minors of* $\hat{\mathbb{H}}$ *of order* k *have the signs of* $(-1)^k$, $k = 2m + 1, \ldots, m + n$, *and where* $F = f + \lambda g$.

Example. Let f and g be as in exercise 3 following section 1. $F = 6 - 4x^1 - 3x^2 + \lambda(1 - (x^1)^2 - (x^2)^2)$. Setting $F_1 = 0$ and $F_2 = 0$ we get

(i) $\quad - 4 - 2\lambda x^1 = 0$

(ii) $\quad - 3 - 2\lambda x^2 = 0$

We know that $x^1 \neq 0$, $x^2 \neq 0$ and $\lambda \neq 0$ or (i) and (ii) will not be satisfied. Thus $x^1 = \frac{4}{3} x^2$. Using the constraint we get $1 = (\frac{25}{g})(x^2)^2$. Thus, by i and ii, we have $(\hat{x}_1; \lambda_1) = (-\frac{4}{5}, \frac{3}{5}; -\frac{5}{2})$ and $(\hat{x}_2; \lambda_2) = (-\frac{4}{5}, -\frac{3}{5}; \frac{5}{2})$ as our critical points and Lagrange multipliers. For $\hat{x}_1$ the bordered Hessian $\hat{\mathbb{H}}^1 = (-1)\begin{bmatrix} 0 & -\frac{8}{5} & -\frac{6}{5} \\ -\frac{8}{5} & 5 & 0 \\ -\frac{6}{5} & 0 & 5 \end{bmatrix}$

$= \begin{bmatrix} 0 & 1.6 & 1.2 \\ 1.6 & -5 & 0 \\ 1.2 & 0 & -5 \end{bmatrix}$ the principal minors of order 3, all we have to look at, are all positive. There is only one and its value is 20. Thus, by Theorem 7 the point $\hat{x}_1$ is a minimizing point for f subject to $g = 0$ (well actually by a corollary to theorem 7). The relevant principal minor of $\hat{\mathbb{H}}^2 = (-1)\begin{bmatrix} 0 & \frac{8}{5} & \frac{6}{5} \\ \frac{8}{5} & -5 & 0 \\ \frac{6}{5} & 0 & -5 \end{bmatrix} = \begin{bmatrix} 0 & -1.6 & -1.2 \\ -1.6 & 5 & 0 \\ -1.2 & 0 & 5 \end{bmatrix}$ has a negative determinant (its value is -20) and so $\hat{x}^2$ is a maximizing point for f subject to $g = 0$. What we did in the example is identify the critical points of the Lagrangian and then use sufficiency theorem to clinch the case against the usual suspects among them.

Exercise 1: Find all the maxima and minima of $f(x) = x^1x^2x^3$ subject to the constraints $g^1(x) = 5 - x^1 - x^2 - x^3 = 0$, $g^2(x) = 8 - x^1x^2 - x^2x^3 - x^3x^1 = 0$.

Exercise 2: Find maxima and minima (all of them) for $f = x^1 - 2x^2 + 2x^3$, $g = (x^1)^2 - (x^2)^2 - (x^3)^2$.

9. NOTES ON THE LITERATURE

The theorems stated in this chapter are common place and have existed for a long time. The original idea for the first order condition can be found in Lagrange (1762). They are stated by way of introduction to earlier works on the problem of Lagrange or Bolza in the calculus of variations, see e.g. Bliss(1930), Caratheodory (1935), Pennisi (1953), and McShane (1941). In Samuelson (1947), the source of the theorems in this chapter is given as Caratheodory (1935) which was not then available in English instead of Bliss (1946) which was. Econometrica published a full length paper, Burger (1955), which consisted of a proof of theorem 7 in this chapter. And Econometrica published a paper, Debreu (1952), providing a characterization of semidefiniteness and definiteness of quadratic forms under linear constraints.

CHAPTER 3

INEQUALITIES AS ADDED CONSTRAINTS

We discuss here the problem of maximizing a real valued function $f: R^n \to R$ subject to equality constraints $g(x) = 0$, where $g: R^n \to R^m$ and subject to inequality constraints $h(x) \geq 0$, where $h: R^n \to R^l$. We refer to the problem as the *Equality-Inequality Constrained Maximization Problem* (EICP). Denoting by C_1 the set $\{x \in R^n | g(x) = 0\}$ and by C_2 the set $\{x \in R^n | h(x) \geq 0\}$, we define a *local solution of EICP* as a point $\hat{x} \in C_1 \cap C_2$ such that $f(\hat{x}) \geq f(x)$ for $x \in N(\hat{x}) \cap C_1 \cap C_2$ where $N(\hat{x}) = R^n$ then $\hat{x}$ is a *global solution of EICP*. In either case we write $\hat{x} = \text{argu max}(f)$ on $N \cap C_1 \cap C_2$.

Formally ECP is equivalent to EICP. That this is clearly so may be seen by noting that an equality constraint can be represented as two inequality constraints, e.g. $h(x) \geq 0$ and $- h(x) \geq 0$, and that an inequality constraint may be represented as an equality constraint by using the Valentine device, see Valentine (1937) of taking up the slack by subtracting the square of a real number and expanding the dimension of the decision space by the number of inequality constraints, so that e.g. $h^\ell(x) \geq 0$ is equivalent to $g^\ell(x,z^\ell) = h^\ell(x) - (z^\ell)^2 = 0$. There is no loss in using the above equivalence if we are only interested in first order necessary conditions. In case of first order sufficient conditions, using EICP as the primary source of theorems restricts the constraints to being linear. On the other hand, any first order sufficiency theorem for ECP leads to such a restriction. In this chapter we derive theorems about EICP from theorems we proved about ECP in the preceding chapter.

1. FIRST ORDER NECESSARY CONDITIONS

We start by introducing some notation and definitions. Let $\hat{x}$ be a point in R^n. We say that h^β is effective at $\hat{x}$ if $h^\beta(\hat{x}) = 0$. For a given $\hat{x}$, let the number of effective constraints be a. We shall renumber the constraints so that the first a components of h are effective. Let h^I be the vector of effective constraints, at $\hat{x}$, and let h^{II} denote the rest of inequality constraints. Thus, $h(\hat{x}) = (h^I(\hat{x}), h^{II}(\hat{x}))$. Finally, let I_z denote a diagonal matrix of order ℓ with z^β's on the diagonal. Similarly define I_{z^I} and $I_{z^{II}}$ to be such matrices of order a and $\ell - a$ respectively.

Theorem 1. *If* 1) f,g *and* h *are of class* C^2, 2) $\hat{x}$ *is a local solution to problem* B, 3) *the matrix* $\begin{pmatrix} \hat{g}_x \\ \hat{h}^I_x \end{pmatrix}$ *has rank* $m + a - n$.

Then there exists a unique vector (ℓ, μ) *such that:*

(a) $\hat{L}_x = 0$, *where* $L = f(x) + \lambda g + \mu h$.

(b) $\mu^\beta h^\beta(\hat{x}) = 0, \quad \beta = 1, ..., \ell$.

(c) $\mu^\beta \geq 0, \quad \beta = 1, ..., \ell$.

Proof: We may rewrite the inequalities $h^\beta \geq 0$ as:

1) $h^\beta(x,z) = h^\beta(x) - (z^\beta)^2 = 0$.

Thus, by hypothesis, $(\hat{x}, \hat{z})$ is a local solution of the following ECP problem: Maximize $f(x)$ subject to $g(x) = 0$, $H(x,z) = 0$. We shall apply theorem 2 of Chapter 2 and we presently show that we may. First we note that the differentiability conditions of theorem 2.2 are satisfied in view of assumption 1 of the present theorem. We now check the rank condition. We must show that the rank $J = \begin{pmatrix} \hat{g}_x & \hat{g}_z \\ \cdots & \cdots \\ \hat{H}^I_x & \hat{H}^I_z \end{pmatrix}$ is $m + \ell$. By hypothesis we have $J^1 = \begin{pmatrix} \hat{g}_x \\ \hat{h}^I_x \end{pmatrix}$ has rank $m + a$. Renumbering the variables so that the first $m + a$ components of x correspond to linearly independent columns of J^1, and partitioning x accordingly into x^I and x^{II} we may write J^1 as $\begin{pmatrix} \hat{g}_{x^I} & \hat{g}_{z^I} \\ \hat{H}^I_{x^I} & \hat{H}^I_{z^I} \end{pmatrix}$. Let $\hat{z}^I$ be the vector of components of $\hat{z}$ corresponding to h^I. We note that $\hat{z}^I = 0$, for $H^I = h^I$ at $\hat{x}$, and that all of the components of $\hat{z}^{II}$ are nonzero. Now $\hat{g}_{z^I} = 0_I$, where 0_I is an $m \times a$ zero matrix and $\hat{g}_{z^{II}} = 0_2$, where 0_2 is an $m \times \ell - a$ zero matrix.

$H_{z^I}^{II} = 0_5$, where 0_5 is an ℓ - a × a zero matrix and $H_{z^{II}}^{II} = -2I_{\hat{z}^{II}}$, where $I_{z^{II}} = I_{z^{II}}|_{z^{II} = \hat{z}^{II}}$. We now write J as:

$$\begin{pmatrix} \hat{g}_{x^I} & \hat{g}_{x^{II}} & \hat{g}_{z^I} & \hat{g}_{z^{II}} \\ \hat{H}^I_{x^I} & \hat{H}^I_{x^{II}} & \hat{H}^I_{z^I} & \hat{H}^I_{z^{II}} \\ \hat{H}^{II}_{x^I} & \hat{H}^{II}_{x^{II}} & \hat{H}^{II}_{z^I} & \hat{H}^{II}_{z^{II}} \end{pmatrix} = \begin{pmatrix} \hat{g}_{x^I} & \hat{g}_{x^{II}} & 0_1 & 0_2 \\ \hat{H}^I_{x^I} & \hat{H}^I_{x^{II}} & 0_3 & 0_4 \\ \hat{H}^{II}_{x^I} & \hat{H}^{II}_{x^{II}} & 0_5 & -2I_{\hat{z}^{II}} \end{pmatrix}$$

It is clear that if $*m + \ell < n$, then J has rank $m + \ell$ since the square submatrix $\begin{pmatrix} \hat{g}_{x^I} & 0_2 \\ \hat{H}^I_{x^I} & 0_4 \\ \hat{H}^{II}_{x^I} & -2I_{\hat{z}^{II}} \end{pmatrix}$ of order $m + \ell$ is non-singular. Now, theorem 2, Chapter 2 applies. We get: There exists a unique vector (λ,μ) such that $\hat{F}_x = 0$ and $\hat{F}_z = 0$, where $F = f(x) + \lambda g + \mu H$. Writing out these conditions we have

$$(1)\ \hat{F}_x = \hat{f}_x + \lambda\hat{g}_x + \mu\hat{h}_x = \hat{L}_x = 0.$$

$$(2)\ \hat{F}_{z^\beta} = -2\mu^\beta\hat{z}^\beta = 0, \quad \text{i.e. } \mu^\beta\hat{z}^\beta = 0.$$

By (1) we have conclusion (a) of the present theorem. (b) follows from (2) since $\hat{z}^\beta \neq 0$ if and only if $h^\beta(\hat{x}) > 0$. We now show that $\mu^I \geq 0$. For the ineffective constraints, h^{II}, $\mu^{II} = 0$. It then remains to show that $\mu^I \geq 0$. In view of the continuity of h^{II}, the components of h^{II} will remain positive in some neighborhood of $\hat{x}$. In that neighborhood $\hat{x}$ is a local solution of the problem: max $f(x,z) = f(x)$ subject to $g(x) = 0$ and $H^I(\hat{x}) = h^I(\hat{x}) = 0$. The conditions for theorem 4 of Chapter 2 (second order necessary conditions) are satisfied. Thus, we have:

$$(3)\ \eta\hat{F}'_{xx}\eta^* + 2\eta\hat{F}'_{xz}\zeta^* + \zeta\hat{F}'_{zz}\zeta^* \leq 0, \text{ where } F' = f + \lambda g + \mu^I H^I,$$

provided:

$$\text{(4.1)}\ \hat{g}_x\eta^* + \hat{g}_z\zeta^* = g_x\eta^* = 0, \text{ and}$$

$$\text{(4.2)}\ \hat{H}^I_x\,\eta^* + \hat{H}^I_z\,\zeta^* = \hat{h}^I_x\,\eta^* = 0, \text{ since } \hat{z}^I = 0.$$

(3) may be written as:

(5) $\eta\hat{F}'_{xx}\,\eta^* - 2\mu^I I_\zeta\zeta^* \leq 0$, with η, ζ satisfying (4), since $\hat{F}'_{xz} = 0$, where I_ζ is a diagonal matrix of order a with ζ's on the diagonal. But (4) does not restrict ζ. Take $(\bar{\eta}; \bar{\zeta})$ to be a vector all of whose components are zero' except for one component of ζ, say $\zeta^{\bar{\beta}}$. $(\bar{\eta}; \bar{\zeta})$ satisfies (4). Thus, (5) holds. Hence $-2\mu^{\bar{\beta}}\zeta^{\bar{\beta}2} \leq 0$. Thus, $\mu^{\bar{\beta}} \geq 0$. Since $\bar{\beta}$ is arbitrary, conclusion (c) of the theorem is proved. This completes the proof of the theorem.

□

Remark: An elegant and simple proof of the above theorem was provided by McShane (1973).

2. FIRST ORDER SUFFICIENT CONDITIONS

In this section we show that, in the case where the equality constraints are linear, if the maximand and inequality constraints are concave, then the vanishing of the first derivatives of the Lagrangian at x is sufficient for x to be a solution to problem B. For the sake of completeness we provide differential characterizations of concavity.

Lemma: *If the real valued* $\phi(x)$ *is of class* C^2 *then the following statements are equivalent for* $x \in R^n$.

(i) $\phi(x)$ *is concave.*

(ii) $\phi(x) - \phi(x^\circ) \leq \phi^\circ_x\xi^*$, $\xi = x - x^\circ$.

(iii) $\xi\phi^\circ_{xx}\xi^* \leq 0$ *for all* ξ *in* R^n.

Remark: In the statement of the lemma, R^n may be replaced by a convex subset of R^n. The proof of the lemma may be found e.g. in Fleming (1977).

Theorem 2. *If* $g(x) = Ax^* + b$, *where* A *is an* $m \times n$ *matrix*, $f(x)$ *and* $h(x)$ *are concave and of class* C^2 *and if:* a) $g(\hat{x}) = 0$, $h(\hat{x}) \geq 0$, b) *There exists a vector* $(\lambda_0, \lambda, \mu)$ *such that:*

(b.i.) $\mu \geq 0$, $\mu g(\hat{x}) = 0$, $\lambda_0 > 0$.

(b.ii.) $\hat{L}^\circ_x = 0$, *where* $L^\circ = \lambda_0 f + \lambda g + \mu h$.

Then $\hat{x}$ *is a global solution of EICP.*

Proof. L° is concave, since λ_g is linear in x, μh is a non-negative linear combination of concave functions and since $\lambda_0 f$ is concave. Thus, by lemma: $L^\circ(x) - L^\circ(\hat{x}) \leq \hat{L}^\circ_x \xi = 0$, for all x. But, $L^\circ(\hat{x}) = \lambda_0 f(\hat{x})$ and for x satisfying the constraints $L^\circ(x) = \lambda_0 f(x) + \mu h(x) \geq \lambda_0 f(x)$. Thus, $\lambda_0(f(x) - f(\hat{x})) \leq L^\circ(x) - L^\circ(\hat{x}) \leq 0$. Since $\lambda_0 > 0$, $f(x) \leq f(\hat{x})$ for all x satisfying $g(x) = 0$ and $h(x) \geq 0$, and the proof is complete.

3. SECOND ORDER NECESSARY CONDITIONS

In this section we utilize theorem 3, Chapter 2 to derive a second order necessary condition for EICP.

Theorem 3. *If the assumptions of theorem 1 are satisfied, then* $\eta \hat{L}_{xx} \eta^* \leq 0$, *for* η *satisfying:* a) $g_x \eta^* = 0$, b) $\hat{h}^I_x \eta^* = 0$, *where* h^I *is the vector of effective constraints at* $\hat{x}$.

Proof. As we did towards the end of the proof of theorem 1 of this chapter, we may express the assumption that $\hat{x}$ is a local solution of problem B as follows: $f(\hat{x}, \hat{z}^I) = f(\hat{x}, 0) = f(\hat{x})$ is a local maximum of $f(x,z) = f(x)$ in some neighborhood of $(\hat{x}, 0)$ satisfying: $g(x,z) = g(x) = 0$, $H^I(x,z) = h^I(x) - I_{z^I} x^{I*} = 0$. As we noted in the proof of theorem 1, theorem 2.3 applies and we have: $\eta \hat{F}'_{xx} - 2\mu^I I_\zeta \zeta^* \leq 0$ provided:

1) $\hat{g}_x\eta^* = 0.$

2) $\hat{h}^I_x\eta^* = 0.$

Since $\eta\hat{F}'_{xx} = \hat{L}_{xx}$ and since 1) and 2) impose no restrictions on ζ, we get:

3) $\eta\hat{L}_{xx}\eta^* - 2\mu^I I_1\zeta^* \leq 0$ for all ζ and for η satisfying 1) and 2).

Hence, in particular for $\zeta = 0$ we have: $\eta\hat{L}_{xx}\eta^* \leq 0$ for η satisfying 1) and 2). This completes our proof. □

4. SECOND ORDER SUFFICIENT CONDITIONS

In this section we use theorem 4, Chapter 2 to obtain a sufficient condition for a point to be a solution of problem B.

Theorem 4. *If* 1) f,g *and* h *are of class* C^2, 2) *the point* $\hat{x}$ *is such that* $g(\hat{x}) = 0$ *and* $h(\hat{x}) \geq 0$, 3) *there exists a vector* $(\lambda_0;\lambda;\mu)$ *such that:* 3.a) $\hat{L}^\circ_x = 0$, *where* $L^\circ = \lambda_0 f + \lambda_g + \mu h$, *and* 3.b) $\mu \geq 0$, $\mu^\beta h^\beta(\hat{x}) = 0$. 4) $\eta\hat{L}_{xx}\eta^* < 0$ for η satisfying $\hat{g}_x\eta^* = 0$, $\hat{h}^I_x\eta^* = 0$. Then $\hat{x}$ is a local solution of EICP.

Proof. We wish to show that there exists a neighborhood of $\hat{x}$, $s(\hat{x})$, such that $f(\hat{x}) - f(x) > 0$ for all x with $g(x) = 0$ and $h(x) \geq 0$. Since $h^{II}(x)$ are continuous and $h^{II}(\hat{x}) > 0$, there exists a neighborhood $T(\hat{x})$ such that $h^{II}(x) > 0$ for $x \in T(\hat{x})$. It suffices to show that there exists a neighborhood $S^1(\hat{x})$ such that $f(\hat{x}) \geq f(x)$ for $x \in S^1(\hat{x})$ satisfying $g(x) = 0$ and $h^I(x) \geq 0$, for we may take $S(\hat{x})$ to be $S(\hat{x}) \cap T(\hat{x})$. As we did in the proofs of theorems 1 and 2, we may write $h^I(x) \geq 0$ as $H^I(x,z) = h^I(x) - z^I I_1 z^{I*} = 0$, where $\hat{z}^I = 0$, and $g(x) = 0$ as $g(x,z) = g(x) = 0$. The proof is complete if we show that there is a neighborhood $S^2(\hat{x},0)$ such that $(\hat{x},0)$ maximizes $f(s,z) = f(x)$ among $(x,z) \in S^2$ satisfying the above equality constraints. To do that we show that theorem 2.4 applies. By assumption 1) of this theorem and by definition of H, assumption 1) of theorem 2.4 is satisfied. Note that $F^\circ = \lambda_0 f + \lambda g + \mu^I H^I = L^\circ - \mu^I I_{z^I} z^{I*}$. Thus, $\hat{L}^\circ_x = 0$ implies $\hat{F}^\circ_x = 0$, since $\hat{z}^I = 0$. Let $Q_0 = (\eta,\zeta)\hat{F}^\circ_{xx}(\eta,\zeta)^*$ and let $Q_1 = \eta\hat{L}^\circ_{xx}\eta^*$. We wish to show that $Q_0 < 0$ for (η,ζ) with:

(i.a) $\hat{g}^I_x\eta^* + \hat{g}_{z^I}\zeta^* = \hat{g}_x\eta^* = 0.$

(i.b) $\hat{H}^I_x\eta^* + \hat{H}^I_z\zeta^* - 2I_{z^I}\zeta^* = \hat{h}^I_z\eta^* = 0.$

By hypothesis, $Q_1 < 0$ subject to (i). But $Q_0 = Q_1 - 2\mu^I I_\zeta\zeta^* \leq Q_1$ for all ζ since $\mu^I \geq 0$ and $\mu I_\zeta\zeta^* = \Sigma\mu^\beta(\zeta^\beta)^2$. Thus, $Q_0 < 0$ subject to (i) and theorem 4, Chapter 2 applies, as we were to show. □

5. NOTES ON THE LITERATURE

The first and second order conditions in this chapter were essentially proved by William Karush in his master's thesis, see Karush (1939). I found that out by reading Pennisi (1953). I got the reference to Pennisi (1953) from Saaty and Bram (1964). Takayama (1974) gives the impression, in p. 61 footnote 5 and p. 73 footnote 7, that Takayama (1974) found out about Karush's contribution. On page 100 of Takayama (1974) the record is set straight by Takayama. When Kuhn in Kuhn (1976) presented a history of the development of mathematical programming he gave credit for the correction of the record to Takayama. Not that any of that is important, but we think it is worthy of note. Professor Kuhn reproduced Karush (1939) as an appendix to Kuhn (1976).

CHAPTER 4

EXTENSIONS AND APPLICATIONS

In this chapter we extend the results of Chapter 3 to deal with the case of vector maxima. But first we relate the results of Chapter 3 to saddle value problems. As we have pointed out, no new results are presented. But we have a unified and, hopefully, more direct treatment of the problems.

In Section 1 we take a special case of the problem of Chapter 3, the case where the constraints are only of the inequality type, and relate the results obtained there to saddle value problems. This results in an alternative proof of Kuhn-Tucker (1951) equivalence theorem.

In Section 2 we extend some results of Chapter 3 and Section 1 to vector maxima. In Section 3 we provide an introduction to discrete time control problems.

Section 4 is devoted to applying theorems 1 and 3 of Chapter 3 to the theory of consumer's optimum. In Section 6 we study the relation between Pareto optimality and competitive equilibrium. In Section 5 we present some preliminary discussion of extensions of the concept of "Lagrangian" and of possible applications to economic theory of the tools developed in part. In Section 7 we revisit the issue of quasi-concavity where the discussion highlights the contribution of Fenchel (1953) to the question of concavifiability. Section 8 presents the usual notes on the literature.

1. RELATION TO SADDLE VALUE PROBLEMS

Here the constraints are the inequality type and they involve non-negativity constraints on the variables. We show, theorem 1, that a local constrained maximum implies the existence of a non-negative saddle point under certain conditions. And conversely, theorem 2, that a non-negative saddle point implies a local constrained maximum. Corollary 1 of theorem 1 and theorem 2 is a restatement of theorem 3 of Kuhn and Tucker (1951). We then relate theorem 2, which is a sufficiency theorem, to the sufficiency of Chapter 3.

Theorem 1 differs from the necessity part of Kuhn and Tucker theorem 3 (1951) in the proof of the quasi-saddle point conditions and in the form of the constraint qualification. We use a form of the constraint qualification presented in the Arrow-Hurwicz-Uzawa paper (1961) on the subject.

For completeness we restate Lemmas 1 and 2 of Kuhn and Tucker together with their sufficient conditions for the hypothesis of lemma 2 of theirs.

We now define a non-negative saddle value of a function. Consider the function $G(x,y)$, where x is a vector with n components and where y is a vector with m.

Definition: A point $(\hat{x},\hat{y}) \geqq 0$ is said to provide a *saddle value* to $G(x,y)$ if:

1.a) $G(x,\hat{y}) \leqq G(\hat{x},\hat{y})$ for all $x \geqq 0$.

1.b) $G(\hat{x},\hat{y}) \leqq G(x,\hat{y})$ for all $y \geqq 0$.

We call $(\hat{x},\hat{y})$ a non-negative *saddle point.*

Let us now present two preliminary lemmas relating saddle values to the derivatives of $G(x,y)$.

Lemma 1). *If*

1) $G(x,y)$ *is differentiable,*

2) $(\hat{x},\hat{y})$ *is a non-negative saddle point of* $G(x,y)$,

then

2.a) $\hat{G}_x \leqq 0$, $\quad \hat{x}^i \hat{G}_{x^i} = 0$; $\quad i = 1, \ldots, n.$

2.b) $G_{y^i} \geqq 0$, $\quad y^i G_{y^i} = 0$; $\quad i = 1, \ldots, m.$

Proof: Let u_i be a vector with m components such that the ith component is 1 and all other components are zeros. Let v_i be a vector with m components all of which are zeros except the ith

which is 1. Let $\Delta G(x_i) = \frac{1}{h}[G(\hat{x} + hu_i, \hat{y}) - G(\hat{x}, \hat{y})]$, where h is a scalar. Similarly let $\Delta G(y_i) = \frac{1}{h}[G(\hat{x}, \hat{y} + hv_i) - G(\hat{x}, \hat{y})]$, where h is a scalar.

Proof of 2.a: Let i_0 be an arbitrary integer between 1 and n. Consider the following two cases:

Case 1: $\hat{x}^{i_0} > 0$. By 1.a, $h = 0$ maximizes the function

$$\omega(h) = G(\hat{x} + hu_{i_0}, \hat{y}), \quad \text{for } x^{i_0} = \hat{x}^{i_0} + hu_{i_0} \geqq 0.$$

Since $\hat{x}^{i_0} > 0$, h may be positive or negative for $x^{i_0} \neq \hat{x}^{i_0}$ and non-negative. Thus:

$$\Delta G(x_i) \leqq 0 \text{ for } h > 0 \text{ and}$$

$$\Delta G(x_i) \geqq 0 \text{ for } h < 0.$$

Passing to the limit as $h \to 0$, we get:

$$\hat{G}_{x^{i_0}} \leqq 0 \text{ and } \hat{G}_{x^{i_0}} \geqq 0.$$

showing, since $\hat{G}_{x^i}$ exists, that $\hat{G}_{x^{i_0}} = 0$. Thus $\hat{G}_{x^{i_0}} \leqq 0$ (with the equality holding) and $\hat{G}_{x^{i_0}}\hat{x}^{i_0} = 0$.

Case 2: $\hat{x}^{i_0} = 0$. Again, by 1.a, $h = 0$ maximizes $\omega(h)$ for $x^{i_0} = \hat{x}^{i_0} + hu_{i_0} \geqq 0$. However, $x^{i_0} = hu_{i_0}$, and $h > 0$ for $0 \leqq x^{i_0} \neq \hat{x}^{i_0}$. Thus:

$$\Delta G(x_i) \leqq 0 \text{ for } x^{i_0} \geqq 0.$$

Now if $x^{i_0} < 0$, either $\omega(h) \geqq \omega(0)$ or $\omega(h) \leqq \omega(0)$. Since h must be negative we have:

$$\Delta G(x_{i_0}) \leqq 0 \text{ or } \Delta G(x_{i_0}) \geqq 0 \text{ for } h < 0.$$

Passing to the limit as $h \to 0$ we get:

$$\hat{G}_{x^{i_0}} \leqq 0 \text{ or } \hat{G}_{x^{i_0}} = 0.$$

In either case $G_{x^{i_0}} \leqq 0$.

This proves 2.a. The proof of 2.b is analogous and the lemma is proved. □

Lemma 2). *If*

1) $g(x,\hat{y})$ *is differentiable,*

2) *condition* 2 *os the conclusion of lemma* 1 *holds at* $(\hat{x},\hat{y})$

and if

3.a) $G(x,\hat{y}) \leqq G(\hat{x},\hat{y}) + \sum_{i=1}^{n} \hat{G}_{x^i}\xi^i,\ \xi^i = x^i - \hat{x}^i,\ x^i \geqq 0,$

3.b) $G(\hat{x},y) \geqq G(\hat{x},\hat{y}) + \sum_{i=1}^{n} \hat{G}_{y^i}\eta^i,\ \eta^i = y^i - \hat{y}^i,\ y^i \geqq 0,$

then $(\hat{x},\hat{y})$ *is a non-negative saddle point of* $G(x,y)$.

Proof: We must show that 1.a and 1.b hold.

1.a):

i) $G(x,\hat{y}) \geqq G(\hat{x},\hat{y}) + \sum_{i=1}^{n} \hat{G}_{x^i}\xi^i = G(\hat{x},\hat{y}) - \hat{G}_{x^i}\hat{x}^i + \sum \hat{G}_{x^i}x^i$ for all

$x^i \geq 0$ by 3.a. But, by 2.a, $\hat{G}_{x^i}\hat{x}^i = 0$ and $\hat{G}_{x^i} \leq 0$.

Thus

ii) $G(x,\hat{y}) - G(\hat{x},\hat{y}) = \sum_{i=1}^{n} \hat{G}_{x^i}x^i \leq 0$ for all $x \geq 0$.

This shows that 1.a holds.

1.b)

By 3.b,

iii) $G(\hat{x},y) \geq G(\hat{x},\hat{y}) + \sum_{i=1}^{m} \hat{G}_{y^i}\eta^i = G(\hat{x},\hat{y}) - \sum_{i=1}^{m} \hat{G}_{y^i}\hat{y}^i + \sum_{i=1}^{m} \hat{G}_{y^i}y^i$ for

$y^i \geq 0$.

But, by 2.b, $\hat{G}_{y^i}\hat{y}^i = 0$ and $\hat{G}_{y^i} \geq 0$, for all $y \geq 0$. Hence $\hat{G}_{y^i}y^i \geq 0$ for $y^i \geq 0$. Thus

$$\text{iv)}\ G(\hat{x},y) - G(\hat{x},\hat{y}) = \sum_{i=1}^{m} \hat{G}_{y^i} \geq 0, \text{ for all } y \geq 0.$$

This shows that relation 1.b follows and the lemma is proved. □

Next we provide sufficient conditions for 3.a and 3.b of lemma 2.

Lemma 3). *If*

1) $G(x,y)$ *is differentiable,*
2) $G(x,y)$ *is concave in* x,
3) $G(x,y)$ *is convex in* y,

Then conditions 3.a *and* 3.b *of lemma* 2 *are satisfied.*

Proof: The proof is due to Kuhn and Tucker (1951) and it involves the definition of convexity, concavity, and the differential. We again repeat that proof for completeness. By definition of concavity, if $x \neq \hat{x}$, we have

$$\theta F(x,\hat{y}) + (1-\theta)G(\hat{x},\hat{y}) \leq \theta \leq 1, \qquad \text{i.e.}$$
$$\theta[G(x,\hat{y}) - G(\hat{x},\hat{y})] \leq G[\hat{x} - \theta(x - \hat{x}),\hat{y}] - G(\hat{x},\hat{y}).$$

Hence, for $0 < \theta \leq 1$

$$G(x,\hat{y}) - G(\hat{x},\hat{y}) \leq \frac{1}{\theta}[G(\hat{x} - \theta(x - \hat{x}),\hat{y}) - G(\hat{x},\hat{y})].$$

By the order preserving property of the limit and by definition of the differential we have

$$G(x,\hat{y}) - G(\hat{x},\hat{y}) \leq \sum_{i=1}^{n} (\hat{G}_{x^i})(x^i - \hat{x}^i) \leq \sum_{i=1}^{n} (\hat{G}_{x^i}\xi^i),$$

which proves that condition 3.a of lemma 2 holds. Analogously we could show that condition 3.b of lemma 2 holds by using the definition of convexity and of the differential. The proof of the lemma is complete. □

Lemma 4). *Let* A *be an open convex subset of* R^n. *And let* f(x) *be of class* C^2 *on* A. f(x) *is concave on* A *if and only if*

$$Q(x,t) = \sum_{i=1}^{n} \sum_{j=1}^{n} f_{ij}(x) t_i t_j$$

is less than or equal to zero for all t *and for all* x *in* A.

Proof: The proof uses the definition of concavity and the mean value theorem. Let x_1 and x_2 be two points in A and let $\bar{x} = \theta x_1 + (1 - \theta) x_2$ for $0 < \theta < 1$. Sufficiency:

By definition of concavity we must show that

i) $\theta f(x_1) + (1 - \theta) f(x_2) - f(\bar{x}) \leq 0$.

But the left hand side of (i) may be written as

$$\theta[f(x_1) - f(\bar{x})] + (1 - \theta)[f(x_2) - f(\bar{x})]$$

By the order mean value theorem there exists a point x' on the segment connecting x_1 and $\bar{x}$ such that

$$f(x_1) - f(\bar{x}) = \sum_i f_i(\bar{x})(x_1^i - \bar{x}^i) + \frac{1}{2} \sum_i \sum_j f_{ij}(x')(x_1^i - \bar{x}^i)(x_1^j - \bar{x}^j) =$$

$$= \sum_i f_i \; f(\bar{x})(1 - \theta)(x_1^i - x_2^i) + \sum_{i,j} f_{ij}(x')(1 - \theta)^2 (x_1^i - x_2^i)$$

So that

$$\text{(ii)} \quad \theta[f(x_1) - f(\bar{x})] = \theta[\sum_i f_i(\bar{x})(1 - \theta)(x_1^i - x_2^i) + \frac{1}{2} \sum_{i,j} f_{ij}(x')(1 - \theta)^2 (x_1^i - x_2^i)(x_1^j - x_2^j)].$$

Similarly, there exists a point x" on the line segment connecting x_2 and $\bar{x}$ such that

(iii) $(1-\theta)[f(x_2)-f(\overline{x})] = (1-\theta)[\sum_i f_i(\overline{x})\theta(x_2^i - x_1^i) +$

$$+\frac{1}{2}\sum_{i,j} f_{ij}(x')\theta^2(x_2^i - x_1^i)(x_2^j - x_1^j)].$$

Adding (ii) and (iii) we get

(iv) $\theta[f(x_1)-f(\overline{x})] + (1-\theta)[f(x_2)-f(\overline{x})] = \theta(1-\theta)\{\sum_i f_i(\overline{x})[(x_1^i - x_2^i)$

$$+(x_2^i - x_1^i)] + \theta Q_1 + (1-\theta)Q_2\} = \theta Q_1 + (1-\theta)Q_2.$$

where Q_1 and Q_2 are the forms in (ii) and (iii) respectively. Now x' and x'' are in A by convexity of A, and hence Q_1 and Q_2 are non-positive. So that $\theta f(x_1) + (1-\theta)f(x_2) - f(\overline{x}) \le 0$. And the sufficiently part of the lemma is proved.

Necessity:

Suppose, by way of contradiction that x_0 is a point in A such that $Q(x,t) = Q(x_0,t_0)$ is positive for some t. Then, by continuity of Q, there exists a neighborhood $N(x_0)$ of x_0 such that $Q(x,t_0)$ is positive for $x \in N(x_0)$. Let $x_2 \in N(x_0)$ be such that $x_2 - x_0 = \lambda t$ for $\lambda \neq 0$. Then we have, by repeating the calculations that lead to (iv),

(v) $\theta f(x_0) + (1-\theta)f(x_2) - f(\overline{x}) = \frac{1}{2}\theta(1-\theta)^2[\sum_{i,j} f_{ij}(x')\lambda^2 t^i t^j] +$

$$+\frac{1}{2}(1-\theta)\theta^2[\sum_{i,j} f_{ij}(x'')\lambda^2 t^i t^j], \text{ where } \overline{x} = \theta x_0 + (1-\theta)x_2,$$

$0 < \theta < 1$, x' is on the line segment connecting $\overline{x}$ and x_0 abd x'' is on the line segment connecting $\overline{x}$ and x_2. Noting that the left hand side of (v) is positive we obtain a contradiction to the concavity of $f(x)$. This proves lemma 4. □

We now have the necessary equipment to develop the theorems of this section.

The problem we are concerned with here, is the maximization of a function $f(x)$, $x \in E^n$, subject to $h^\beta(x) \geq 0$, $\beta = 1, \ldots, \ell$ and to $x \geq 0$. Ths the local constrained maximum at x_0 means that there exists a neighborhood of x_0 such that $f(x_0) \geq f(x)$ for all $x \geq 0$ in that neighborhood satisfying $h^\beta(x) \geq 0$, $\beta = 1, \ldots, \ell$.

Theorem 1. *If* 1) $f(x_0)$ *is a local constrained maximum,* 2) *the function* f *and* h^β *have continuous second order derivatives,* 3) *the rank of Jacobian of effective constraints* h^β, *at* $\hat{x}$ *evaluated at* $\hat{x}$ *is equal to the number of effective constraints,* 4) *the number of effective constraints plus the number of zero components of* $\hat{x}$ *is less than* n, 5) f *and* h^β *are concave in* x, *then there exists a vector* $\hat{\mu} = (\hat{\mu}^1, \ldots, \hat{\mu}^\ell)$ *with* $\hat{\mu}^\beta \geq 0$, $\beta = 1, \ldots, \ell$, *such that* $L(\hat{x},\hat{\mu})$ *is a non-negative saddle value of* $L(x,\mu) = f(x) + \sum \mu^\beta h^\beta$.

Proof: The proof of the theorem consists of verifying the conditions of lemma 1 of this section. This, in turn, is done by showing that we may apply theorem 1 of Chapter 3 to our present problem.

Let $h^{\ell+i}(x) = x^i$, $i = 1, \ldots, n$. Then by hypothesis $f(x_0)$ is a local maximum of $f(x)$ subject $h^\beta \geq 0$ and $h^{\ell+i}(x) \geq 0$, $\beta = 1, \ldots, \ell$; $i = 1, \ldots, n$. Let $h^{(1)}(x)$ be the vector whose components are the h^β's and let $h^{(2)}(x)$ be the vector whose components are the h^{m+i}'s. Let $h^{\Gamma_i}(x)$ be the vector whose components are the zero components of $h^{(i)}(\hat{x})$, $i = 1, 2$. Let $h^\Gamma(x) = (h^{\Gamma_1}, h^{\Gamma_2})$. The matrix $h^\Gamma_x(x)$ has the form $\begin{bmatrix} h^{\Gamma_1}(\hat{x}) \\ h^{\Gamma_2}(\hat{x}) \end{bmatrix}$ where the component $h^i_j(\hat{x})$ of $h^{\Gamma_2}_x(\hat{x})$ is 1 when $i = j$ and zero otherwise. One may, by renumbering the variables, choose a square submatrix of $h^\Gamma_x(\hat{x})$ of order equal to the number of rows that matrix that is non-singular. Thus the rank condition of theorem 1 of Chapter 3 is satisfied. The other conditions of that theorem are satisfied by virtue of the hypothesis of the present theorem. Thus there exists a vector of multipliers $\bar{\mu} = (\bar{\mu}^1, \bar{\mu}^2) \geq 0$ such that

(i) $L_{x^i}(\hat{x},\bar{\mu}) = 0$, where $L(\hat{x},\bar{\mu}) = f(x) + \sum_{\beta=1}^{\ell} \bar{\mu}^{\beta} h^{\beta}(x) + \sum_{i=1}^{n} \bar{\mu}^{\ell+i} h^{\ell+i}(x)$.

(ii-a) $\bar{\mu}^{\beta} h^{\beta}(\hat{x}) = 0, \ \beta = 1, \ldots, \ell$

(ii-b) $\bar{\mu}^{\ell+1} h^{\ell+i}(\hat{x}) = 0, \ i = 1, \ldots, n.$

Now

$$L_{x^i}(\hat{x},\bar{\mu}) = f_i(\hat{x}) + \sum_{\beta=1}^{\ell} \bar{\mu}^{\beta} h_i^{\beta}(\hat{x}) + \bar{\mu}^{\ell+1} = 0.$$

Take $\hat{\mu} = \bar{\mu}^1$. Then $L_{x^i}(\hat{x},\bar{\mu}) = L_i(\hat{x},\hat{\mu}) + \bar{\mu}^{\ell+i} = 0$. But $\bar{\mu}^{\ell+i} \geq 0$ and so

(iii) $L_{x^i}(\hat{x},\hat{\mu}) \leq 0.$

Also, by (ii-b), the equality holds if $\hat{x}^i \geq 0$, i.e.

(iv) $\hat{L}_{x^i} \leq 0, \qquad \hat{x}^i \hat{L}_{x^i} = 0, \qquad i = 1, \ldots, n.$

Noting that $\hat{x} \geq 0$ and $h^{\beta}(\hat{x}) \geq 0$, by hypothesis of this theorem, we have

(v) $\hat{L}_{\mu^{\beta}}(\hat{x},\hat{\mu}) = h^{\beta}(\hat{x}) \geq 0, \qquad \beta = 1, \ldots, \ell.$

Furthermore, by (ii-a)

(vi) $\hat{\mu}^{\beta} h^{\beta}(\hat{x}) = 0.$

By (v) and (vi) we have

(vii) $\hat{L}_{\mu^{\beta}} \geq 0, \qquad \hat{\mu}^{\beta} h^{\beta}(\hat{x}) = 0, \qquad \beta = 1, \ldots, \ell.$

We may now apply lemma 2 above to the function $L(x,\mu)$. By (iv), (vii), the fact that $(\hat{x},\hat{\mu}) \geq 0$, assumption 5 of this theorem, concavity of $L(x,\mu)$ in x and linearity - hence convexity - of $L(x,\mu)$ in μ the lemma applies. □

Theorem 2. *If there exists an m-dimensional non-negative vector* μ_0 *such that* $L(x_0,\mu_0)$ *is a non-negative saddle value of* $L(x,\mu)$, *where* $L(x,\mu)$ *is as in theorem 1, then* $f(\hat{x})$ *is a constrained maximum in the sense of this section.*

Proof: The proof consists of simply applying the definition of non-negative saddle point to the function $L(x,\mu)$. Doing so, we have

$$L(x,\hat{\mu}) \leq L(\hat{x},\hat{\mu}), \qquad x \geq 0, \text{ i.e.}$$

(i) $$f(x) + \sum_{\beta} \hat{\mu}^{\beta} h^{\beta}(x) \leq f(\hat{x}) + \sum \hat{\mu}^{\beta} h^{\beta}(\hat{x}), \qquad x \geqq 0.$$

Since $L(\hat{x},\hat{\mu})$ is a saddle value for $L(x,\mu)$ we have, by lemma 1,

(ii) $\hat{\mu}^{\beta} h^{\beta}(\hat{x}) = 0.$

Thus, by (i), $f(x) + \sum_{\beta} \hat{\mu}^{\beta} h^{\beta}(x) \leq f(\hat{x})$, i.e.

(iii) $$f(\hat{x}) - f(x) \geqq \sum_{\beta} \hat{\mu}^{\beta} h^{\beta}(x), \; x \geqq 0.$$

Noting that the right hand side of the inequality (iii) is non-negative for any x satisfying $h^{\beta}(x) \geq 0$, $\beta = 1, \ldots, \ell$. □

Corollary 1) *f conditions 2, 3, 4, and 5 of theorem 1 are satisfied then* $f(\hat{x})$ *is a local constrained maximum if and only if there exists* $\hat{\mu} \geqq 0$ *such that* $L(\hat{x},\hat{\mu})$ *is a non-negative saddle value for* $L(x,\mu)$.

Proof. The sufficiency part follows directly from theorem 2 and the necessity part follows, also directly from theorem 1.

Remark 1) Let us define a local non-negative saddle point as a non-negative point that satisfies inequalities 1.a) for $x \geq 0$ in a neighborhood of $\hat{x}$ and similarly with 1.b. Theorems 1 and 2 and Corollary 1 hold if we replace "non-negative saddle point" by "local non-negative saddle point".

Remark 2) Corollary 1 is quite expensive for, as is clear from theorem 2, all that is needed for the sufficiency part is that $L(\hat{x},\hat{\mu})$ be a non-negative saddle point.

Theorem 3. *If* 1) $f(x)$ *and* $h^\beta(x)$, $\beta = 1, ..., \ell$, *have continuous second order derivatives,* 2) *there exists a vector* $\mu \geq 0$ *such that* $\hat{L}_{x_i} = 0$, $i = 1, ..., n$, *where* $\hat{L}(x,\bar{\mu})$ *is as defined in the proof of theorem 1,* 3) $Q(\hat{x},\eta) = \eta\hat{L}_{xx}\eta^* < 0$ *for* $\eta \neq 0$. 4) $h^\beta(\hat{x}) \geq 0$, $\beta = 1, ..., \ell$, *then there exists a neighborhood of* $\hat{x}$, *such that* $f(\hat{x})$ *is a local constrained maximum.*

First proof of theorem 3: Clearly condition 3) implies that condition 4) of theorem 4 of Chapter 3 is satisfied. Furthermore, all other condition of that theorem hold and the present theorem follows.

Second proof: By an argument similar to that used in the proof of theorem 1 we have relations (iv) and (vii) in that proof. By assumption 3) of this theorem, note that by continuity of the elements of L_{xx} there exists an open sphere with center at $\hat{x}$ such that $\eta F_{xx}\eta^* < 0$ for x in that sphere. Thus by lemma 4, $L(x,\mu)$ is concave in x in that space.

Finally, by linearity in μ, $L(x,\mu)$ is convex in μ. Thus lemma 2 applies, locally, and $L(\hat{x},\hat{\mu})$ is a local non-negative saddle point of $L(x,m)$. Now the conclusion follows from theorem 2 if we confine our attention to the sphere mentioned above.

Remark 3) The two ways of proving theorem 3 indicate the relation between the two types of sufficiency theorems but do not give their exact relationship. They show that sufficient conditions for a local non-negative saddle point imply the sufficient conditions for a local constrained maximum.

2. EXTENSION TO VECTOR MAXIMA

Let $F(x)$ be a k-dimensional vector valued function of x, $F(x) = (f^1(x), ..., f^k(x))$.

Definition 1) $F(\hat{x})$ is said to be a constrained vector maximum of $F(x)$ - subject to the constraint $h^\beta(\bar{x}) \geq 0$, $\beta = 1, ..., \ell$, $x \geq 0$ - if there does not exist any $\bar{x} \geq 0$ with $h^\beta(\bar{x}) \geq 0$, $\beta = 1, ..., \ell$, such that $f^j(\bar{x}) \geq f^j(\hat{x})$ with strict inequality for at least one j.

Remark: It follows from the above definition that if $F(\hat{x})$ is a constrained vector maximum, then $f^{j_0}(\hat{x})$ is a maximum subject to: $f^j(x) \geq f^j(\hat{x})$, $j \neq j_0$, between 1 and k, and to: $h^\beta(x) \geq 0$, $\beta = 1, ..., \ell$; $x \geq 0$, for $j_0 = 1, .., n$. For if not, then there exists an $\bar{x} \geq 0$ and a $\bar{j}$ such that:

$$f^{\bar{j}}(\bar{x}) \geq f^{\bar{j}}(\hat{x})$$

$$f^j(\bar{x}) \geq f^j(\hat{x}), \qquad i \neq \bar{\imath} \text{ between } 1 \text{ and } n,$$

$$h^\beta(\bar{x}) \geq 0.$$

Utilizing the remark we provide a calculus proof of a theorem that is closely related to theorem 4 of Kuhn and Tuckner (1951). The method of the proof was suggested by Professor L. Hurwicz.

Definition 2) Fix j_0. A point $\hat{x}$ is said to be *j_0-regular* of the matrix

$$\begin{bmatrix} \hat{F}^{\Gamma}_x \\ \hat{h}^{\Gamma}_x \end{bmatrix}$$

has maximum rank, where $F^{\Gamma}(x)$ is a vector whose components are f^j's with $j \neq j_0$, $h^{\Gamma}(x)$ is the vector whose components are $h^{\beta}(x)$ such that $h^{\beta}(\hat{x}) = 0$.

Definition 3) A point $\hat{x}$ is said to be *regular* if it is j_0-regular for $j_0 = 1, ..., n$.

Definition 4) $\hat{x}$ is said to by j_0-constrained maximum if statement (I) holds.

Theorem 4. *If* 1) $F(\hat{x})$ *is a constrained vector maximum,* 2) $\hat{x}$ *is regular,* 3) $f^j(x)$ *and* $h(x)$ *have continuous second order partial derivatives,* 4) ℓ + *the number of effective constraints, including* $x^i \geq 0$, *is less than* n, *then there exist constants* $\hat{v}^j > 0, j = 1, ..., n$, *and a m-vector* $\hat{\mu} \geq 0$ *such that*

(i) $\hat{L}_{x^i} \leq 0, \qquad \hat{x}^i \hat{L}_{x^i} = 0, \qquad i = 1, ..., n,$

(ii) $\hat{L}_{\mu^\beta} \geq 0, \qquad \hat{\mu}^\beta \hat{L}_{\mu^\beta} = 0, \qquad \beta = 1, ..., \ell,$

where $L(\mu,x) = \psi(x) + \sum_{\beta=1}^{\ell} \mu^\beta h^\beta(x)$, *and* $\psi(x) = \sum_{j=1}^{\ell} \hat{v}^j f^j(x)$.

Proof: By the remark at the beginning of this section, $\hat{x}j_0$-constrained maximum for $j_0 = 1, ..., n$. Hence $\hat{x}$ is a locally j_0-constrained maximum, i.e. $f^{j_0}(\hat{x})$ is a local maximum subject to:

$$h^\beta(x) \geq 0$$
$$h^{m+j}(x) = f^j(\hat{x}) \geq 0, \qquad j \neq j_0$$
$$h^{m+(\ell-1)+i}(x) = x^i \geq 0.$$

Fix j_0. Since x is a j_0-regular, the rank condition in theorem 1 of Chapter 2 is satisfied, by an argument similar to that of the proof of theorem 1 of this chapter. We get, again using the same argument we used improving relations (iv) and (vii) in the proof o theorem 1,

there exists vectors $\lambda^{j_0} \geq 0$ and $\mu^{j_0} \geq 0$, where the $j_0{}^{th}$ component of the ℓ-dimensional vector $\lambda_0^{j_0}$ is equal to one, such that

(i) $\hat{L}^{j_0}_{x_i} \leqq 0, \qquad x_0^i \hat{L}^{j_0}_{x_i} = 0, \qquad i = 1, ..., n,$

(ii) $\hat{L}^{j_0}_{\lambda^{j_0}} \leqq 0, \qquad \lambda_0^{j_0} \hat{L}^{j_0}_{j_0} = 0, \qquad j \neq j_0,$

(iii) $\hat{L}^{j_0}_{\mu^{j_0,\beta}} \geqq 0 \qquad \mu_0^{j_0,\beta} \hat{L}^{j_0}_{j_0,\beta} = 0, \qquad \beta = 1, ..., \ell,$

where

$$L^{j_0}(\lambda^{j_0},\mu^{j_0},x) = f^{j_0}(x) + \sum_{j \neq j_0} \lambda^{j_0,j}[f^j(x) - f^j(x_0)] + \sum_{\beta} \mu^{j_0,\beta} h^{\beta}(x).$$

Repeating the above argument as j_0 goes from 1 to k we have for $j_0 = 1, ..., \ell$

(iv) $\hat{L}^{j_0}_{x_i} = \hat{f}^{j_0}_{x_i} + \sum_{j \neq j_0} \lambda^{j_0,j} \hat{f}^j_{x_i} + \sum_{\beta} \mu^{j_0,\beta} \hat{h}^{\beta}_{x^i} \leqq 0, \quad i = 1, ..., n.$

(v) $\hat{L}^{j_0}_{\lambda^{j_0,j}} = f^j(\hat{x}) - f^j(\hat{x}) = 0, \qquad j \neq j_0,$

(vi) $\hat{L}^{j_0}_{\mu^{j_0,\beta}} = h^{\beta}(x) \geq 0, \qquad \mu^{j_0,\beta} \hat{L}^{j_0}_{\mu^{j_0,\beta}} = 0, \qquad \beta = 1, ..., m.$

Summing each of systems (iv) and (vi) over j_0 we get:

(vii) $\sum_{j=1}^{\ell} (1 + \sum_{j \neq j_0} \lambda^{j_0,j}) \hat{f}^j_{x^i} + \sum_{\beta=1}^{m} (\sum_{j_0=1} \mu^{j_0,\beta}) \hat{h}^{\beta}_{x^i} \leqq 0$, with equality if $\hat{x}^i > 0$,

i = 1,

..., n.

(viii) $\ell h^{\beta}(\hat{x}) \geqq 0, \quad (\sum_{j_0=j}^{\ell} \mu^{j_0,\beta}) h(\hat{x}) = 0, \qquad \beta = 1, ..., \ell$, i.e.

(ix) $h^{\beta}(\hat{x}) \geq 0, \quad (\sum_{j_0=1}^{\ell} \mu^{j_0,\beta}) h^{\beta}(\hat{x}) = 0.$

Now let $\hat{v}^j = 1 + \sum_{j \neq j_0} \lambda_0^{j_0,j}$ and let $\hat{\mu} = \sum_{j_0=1}^{\ell} \mu^{j_0,\beta}$. Clearly $\hat{v}^j > 0$ for $j = 1, \ldots, \ell$,

and $\hat{\mu}^{\beta} \geq 0$ for $\beta = 1, \ldots, m$. Consider

$$L(x,\mu) = g(x) + \sum_{\beta=1}^{\ell} \mu^{\beta} h^{\beta}(x), \text{ where } g(x) = \sum_{j=1}^{\ell} \hat{v}^j f^j(x).$$

The conclusion of the theorem follows from (vii) and (ix) above. □

Corollary 1) *If, in addition to the hypothesis of theorem 4, the functions* $f^j(x)$ *and* $h^{\beta}(x)$ *are concave,* $j = 1, \ldots, \ell$; $\beta = 1, \ldots, m$, *then* $(\hat{x},\hat{\mu})$ *is a non-negative saddle point for* $\psi(x,\mu)$.

Proof: The corollary follows by combining the results of theorem 4 and lemma 2 of Section 1 taking into consideration that $L(x,\mu)$ is linear, hence convex, in μ.

3. A DISCRETE TIME OPTIMAL CONTROL PROBLEM

Consider a control system where the equations of motion are given by the system of difference equations:

(1) $x^{t+1} - x^t = f^t(x^t,u^t), \qquad t = 1, \ldots, T - 1,$

where x^t is an n-vector and where u^t is an m-vector. Let x denote the finite sequence $\{x_1, ..., x_T\}$ and let u denote the finite sequence $\{u_1, ..., u_T\}$. We shall consider two problems:

Problem I: Maximize J(x,u) subject to (1) and to

(2.1) $\bar{h}^t(x^t,u^t) = 0, \qquad 1 \le t \le T - 1$

(2.2) $\bar{\bar{h}}^t(x^t,u^t) \ge 0, \qquad 1 \le t \le T - 1$, where $\bar{h}$ is an N_1-vector function and $\bar{\bar{h}}$ is an N_2-vector function.

(3.1) $\bar{g}(x^1,x^T) = 0,$

(3.2) $\bar{\bar{g}}(x^1,x^T) \ge 0$, where $\bar{g}$ is an M_1-vector function and $\bar{\bar{g}}$ is an M_2-vector function.

Problem II: Maximize $\sum_{t=1}^{T-1} R^t(x^t,u^t) + R^T(x^T)$ subject to (1), (2), and (3).

Problem II is a special case of problem I, and is the more familiar form of the optimal control problem. We shall only provide a brief introduction to the problem. The reader is referred to Canon, Cullum, and Polak (1970) for more detailed and general treatment of the problem and for references.

First we characterize solutions to problem I and then we use the results to characterize problem II. Let $\overset{\circ}{z} = (\overset{\circ}{x},\overset{\circ}{u})$ be a given vector in the $nT + m(T - 1)$-dimensional Euclidean space.

Proposition 1: *If* 1) J, f, h, *and* g *are of class* C^1. 2) z *is a solution to problem I. Then there exists vectors* $(\lambda_0,\bar{\gamma},\bar{\bar{\gamma}})$ *and sequences* $\{\lambda^t,\bar{\gamma}^t,\bar{\bar{\gamma}}^t\}$ *with* $(\lambda_0, \gamma, \lambda, \mu) \ne 0, \lambda_0 \ge 0$, *such that:*

(4.1) $\bar{\bar{\gamma}} \ge 0, \qquad \overline{\overline{\gamma g}}(\hat{z}) = 0$

(4.2) $\bar{\bar{\mu}} \ge 0, \qquad \bar{\bar{\mu}}^t\bar{\bar{h}}^t(\hat{z}) = 0, \ t = 1, ..., T - 1.$

(5) $\lambda^{t-1} - \lambda^t = \lambda_0\hat{J}_{x^t} + \lambda^t\hat{f}^t_{x^t} + \mu^t\hat{h}^t_{x^t}, \qquad t = 2, ..., T - 1.$

(6) $\lambda_0\hat{J}_{u^t} + \lambda^t\hat{f}^t_{u^t} + \mu^t\hat{h}^t_{u^t} = 0, \qquad t = 1, ..., T - 1.$

(7.1) $\lambda_0\hat{J}_{x^1} + \lambda^1 + \lambda^1\hat{f}^1_{x^1} + \mu^1\hat{h}^1_{x^1} + \gamma\hat{g}_{x^1} = 0.$

(7.2) $\lambda_0\hat{J}_{x^T} + \gamma\hat{g}_{x^T} = 0.$

Proposition 2. *If* 1) J, f, h *and* g *are of class* C^1. 2) f, $\bar{h}$, and $\bar{\bar{g}}$ *are linear.* 3) J, $\bar{\bar{h}}$, $\bar{\bar{g}}$ *are concave.* 4) *Conditions (4)-(6) of proposition 1 hold at a point* z *which satisfies constraints (1)-(3). Then* z *is a solution to problem 1.*

Outline of proofs:

To prove proposition 1, apply theorem 1 of Chapter 1. Define $F^I = J + \sum_{t=1}^{T-1} \lambda^t(f^t(x^t,u^t) - x^{t+1} + x^t) + \sum_{t=1}^{T-1} \mu^t h^t + \gamma g$. The existence of the multipliers $\lambda_0, \lambda, \mu, \gamma$ is simplified by the theorems and they satisfy properties (4) of proposition 1. Writing down the fact that the derivatives of F with respect to x^t and u^t and equating the expressions at z to zero we have:

(8) $\hat{F}^I_{u^t} = \lambda_0\hat{J}_{u^t} + \lambda^t\hat{f}^t_{u^t} + \mu^t\hat{h}^t_{u^t} = 0, \qquad t = 1, ..., T - 1.$

(9) $\hat{F}^I_{x^1} = \lambda_0\hat{J}_{x^1} + \lambda^1 + \lambda^1\hat{f}^1_{x^1} + \mu^1\hat{h}^1_{x^1} + \gamma\hat{g}_{x^1} = 0$

(10) $\hat{F}^I_{x^t} = \lambda_0\hat{J}_{x^t} + \lambda^t - \lambda^{t-1} + \lambda^t\hat{f}^t_{x^t} + \mu^t h^t_{x^t} = 0, \qquad t = 1, ..., T - 1.$

(11) $\hat{F}^I_{x^t} = \lambda_0\hat{J}_{x^t} + \hat{g}_{x^T} = 0.$

Conditions (10), (8), (9), and (11) are equivalent to conditions (5), (6), (7.1), and (7.2), respectively, of proposition 1.

To prove proposition 2, we observe that theorem 3 of Chapter 1 applies, in view of (8) - (11), and in view of assumptions (2) and (3) of the proposition.

We are now ready to characterize solutions to problem II.

Proposition 3. *If* 1) R, f, h, *and* g *are of class* C^1. 2) $\hat{z}$ *is a solution to problem II. Then there exists a vector* (λ_0, γ) *and sequences* $\{\lambda^t\}$ *and* $\{\mu^t\}$ *with* $\{\lambda_0, \gamma, \lambda^t, \mu^t\} \neq 0$ for $t = 1, ..., T - 1$, *such that, in addition to (4) in proposition 1, we have:*

(12) $\lambda^{t-1} - \lambda^t = \hat{H}^t_{x^t}$, $\quad t = 2, ..., T - 1$, *where* $H^t = \lambda_0 R^t + \lambda^t f^t + \mu^t h^t$,

(13) $\hat{H}_{u^t} = 0$, $\quad t = 1, ..., T$.

(14.1) $\quad \hat{G}_{x^1} + \hat{H}^1_{x^1} + \lambda^1 = 0$, *where* $G = \lambda_0 R^1 + \lambda_0 R^T + \gamma g$.

(14.2) $\quad \hat{G}_{x^T} = 0$.

Proposition 4) *If* 1) R, f, h, *and* g *are of class* C^1. 2) $\hat{z}$ *satisfies constraints (1)-(3).* 3) f, $\bar{h}$, *and* $\bar{g}$ *are linear.* 4) J, $\bar{\bar{h}}$, $\bar{\bar{g}}$ *are concave.* 5) *Conditions (4), (12)-(14) are satisfied at* $\hat{z}$. *Then* $\hat{z}$ *is a solution to problem II.*

Propositions 3) and 4) follow immediately from propositions 1) and 2). Compare propositions 3) and 4) to the results in section 1 of Chapter 7.

4. CONSUMER'S OPTIMUM, ONCE MORE

Let $x = (x_1, ..., x_n)$ denote a commodity bundle. We speak of a commodity space as the set of all commodity bundles and denote it by $\mathcal{X}$. We define a consumer as an entity with: a) A complete ordering of the commodity space and b) An income, denoted by m. Let $p = (p_1, ..., p_n)$ denote a price vector. The expenditure of the consumer, if he buys a bundle x, is $p \cdot x$, where $p \cdot x = \sum_{i=1}^{n} p_i x_i$. We postulate that the consumer chooses the bundle which is most preferred subject to the budget constraint $p \cdot x \leq m$ and provided that x lies in a subset of

$\mathcal{X}$ which we call the consumption set and denote by $\mathcal{X}^*$. We shall assume, further that:

A.1) $\mathcal{X}$ is the Euclidean n-space.

A.2) The consumer's ordering is representable by a real valued function U(x).

A.3) That $\mathcal{X}^*$ is the non-negative orthant of $\mathcal{X}$, i.e. $\mathcal{X}^* = \{x \in \mathcal{X}: x \geq 0\}$.

We now present some of the well known propositions about consumer's optimum. We start by introducing some lemmas which indicate the characterization of a consumer's optimum.

Lemma 1: *(first order necessary conditions) If*

1) *The function* U *is continuously differentiable,*
2) *The consumer takes prices parametrically, i.e.* p *is independent from* x,
3) $\hat{x}$ *is a consumer's optimum (i.e.* $\hat{x}$ *maximizes* U(x) *subject to* $m - p \cdot x \geq 0$ *and* $x \geq 0$),
4) *Not all prices are zero (i.e.* $p \neq 0$*).*

Then there exist a positive constant λ_0 *and a non-negative constant* μ *such that:*

1) $\lambda_0 \hat{U}_i \leq \mu p^i$ *with equality if* $\hat{x}_i > 0$, $(i = 1, ..., n)$,

2) $\mu(m - p \cdot \hat{x}) = 0$, *where* $\hat{U}_i = \left.\frac{\partial U}{\partial x^i}\right|_{x = \hat{x}}$ *and* p^i *is the* i^{th} *component of* p.

Proof: $\hat{x}$ is a solution to the following forms: max U(x) subject to $h^1(x) = m - p \cdot x \geq 0$, $h^{1+i}(x) = x^i \geq 0$, $i = 1, ..., n$. Theorem 1 of section 1 applies (if the budget constraint is effective at $\hat{x}$, then condition 4 of the lemma guarantees that the rank condition is satisfied). Thus there exist a positive constant λ_0 and non-negative constants $\mu, \gamma^1, ..., \gamma^n$ such that: $\hat{F}_i = 0 (i = 1, ..., n)$, where $F = \lambda_0 U(x) + \mu(m - p \cdot x) + \sum_{i=1}^{n} \gamma^i x^i$ and $\hat{F}_i = \left.\frac{\partial F}{\partial x^i}\right|_{x = \hat{x}}$,

b) $\mu(m - p \cdot \hat{x}) = 0$, $\gamma^i \hat{x}^i = 0$. From a) and b) it follows that: c) $\lambda_0 \hat{U}_i - \mu p^i + \gamma^i = 0$, $\gamma^i \hat{x}^i = 0$, and d) $\mu(m - p \cdot x) = 0$. d) establishes conclusion 2) of the lemma. From c) we have $\lambda_0 U_i - \mu p^i = -\gamma^i$, $\gamma^i x^i = 0$. But $-\gamma^i \leq 0$ and conclusion 1) of the theorem is proved. □

Lemma 2: *(second order necessary conditions) If assumption 1) of lemma 1 is replaced by 1)' the function* U *has continuous second order derivatives. And if assumptions 2) and 4) of lemma 1 hold, then there exist* λ_0 *and* μ *as in lemma 1 such that* $\sum_{i,j} \hat{U}_{ij} \eta^i \eta^j \leq 0$ *for all* $\eta \neq 0$ *satisfying,*

(i) $p \cdot \eta = 0$ *whenever* $m = p \cdot \hat{x}$

(ii) $\eta^i = 0$ *whenever* $\hat{x}^i = 0$,

where $U_{ij} = \left. \dfrac{\partial^2 U}{\partial x^i \partial x^j} \right|_{x = \hat{x}}$ *and where* η^i *and* η^j *denote components of* η.

Proof: Theorem 3 of Chapter 3 applies. Thus, with F as in the proof of lemma 1, $F_{ij} = U_{ij}$. The conditions on η (given by (i) and (ii)) follow by differentiating h^1 and h^{1+i} with respect to x (where h^1, h^{1+i} are as in the proof of lemma 1). We then have $\lambda_0 \sum_{i,j} U_{ij} \eta^i \eta^j \leq 0$ for $\eta \neq 0$ satisfying (i) and (ii). Since $\lambda > 0$ our conclusion follows. □

We define a demand function as a function $H\colon R_+^{n+1} \to R_+^n$, where R_+^{n+1} is the non-negative orthant of the Euclidian (n+1)-space whose elements are prices and income (p,m) and where R_+^n is the non-negative orthant of the Euclidian n-space whose elements are commodity bundles x. We write demand functions as $x^i = H^i(p,m)$. Suppose H^i to be differentiable. Then we define a Slutsky term as a change in the demand for commodity i due to "compensated" change in the price of commodity j. We take compensation to mean a change in m such that the consumer could afford his previous optimal bundle. Formally, denoting Slutsky terms by K_{ij}, we write:

$$(3) \quad K_{ij} = \left.\frac{\partial H^i}{\partial p_j}\right|_{p.dx=0} = H^i_j + x^j H^i m$$

<u>Proposition 1)</u> *If* 1) *The function* U *has continuous second order derivatives.* 2) *The demand functions are solutions tot he consumer's optimum problems with* $p \cdot x = m$ *and* $x^i > 0$. 3) H^i *are continuously differentiable.* 4) *Not all prices are zero. Then the demand functions have the following properties:*

(i) H^i *are homogeneous of degree zero in prices and income,*

(ii) H^i *satisfy* $p \cdot H = m$,

(iii) *The Slutsky form* $\sum_{i,j} K_{ij} dp^i dp^j \leq 0$.

Proof: Since the demand functions are obtained as solutions to the consumer's optimum problem, by lemma 1 and assumption 3, we have:

$$(4) \quad U_i = \bar{\mu} p_i, \text{ where } \bar{\mu} = \mu/\lambda_0$$

$$(5) \quad p \cdot x = m.$$

Properties (i) and (ii) follow from (5) which must be satisfied at all p and m. To show that (iii) holds we first write:

$$(6) \quad dx^i = dH^i = \sum_j H^i_j dp^j + H^i_m dm.$$

Differentiating (5) we have:

$$(7) \quad \sum_j p^j dx^j + \sum_j x^j dp^j = dm.$$

Taking compensation, $\sum_j p^j dx^j = 0$, into consideration we have:

(7)' $dm = \sum_j x^j dp^j.$

Substituting from (7)' in (6) we have:

(8) $dx^i = \sum_j H^i_j dp^j + H^i_m \sum_j x^j dp^j.$

Multiplying each of the equations (8) by dp^i and summing we get:

(9) $\sum_i dx^i dp^i = \sum_{i,j} (H^i_j + H^i_m x^j) dp_i dp_j = \sum_{i,j} dp^i dp^j.$

Thus, the property (iii) is demonstrated once we show that $\sum_i dx^i dp^i \leq 0$ for $p^i dx^i = 0$. Differentiating (4) we have:

(10) $\sum_j U_{ij} dx^j = \bar{\mu} dp^i + p^i d\mu.$

Multiplying equations (10) by dx^i and adding the resulting equations we have:

(11) $\sum_{i,j} U_{ij} dx^i dx^j = \bar{\mu} \sum_i dp^i dx^i + d\mu \sum p^i dx^i.$

Taking note of the compensation, the last term on the right hand of (11) is zero and

(12) $\bar{\mu} \sum_i dp^i dx^i = \sum_{i,j} U_{ij} dx^i dx^j.$

By lemma 2, since $\sum_i p^i dx^i = 0$, the right hand side of (12) is non-positive. Since $\bar{\mu} \geq 0$ we have: $\sum_i dx^i dp^i \leq 0$, and property (iii) holds.□

<u>Remark 1</u>: If we assume that the matrix $[U_{ij}]$ has an inverse, then it follows from (8), (10), and the symmetry of $[U_{ij}]$ (which follows from assumption 1) of proposition 1) that $[K_{ij}]$ is symmetric.

Remark 2: The proposition is true if we replace the utility index $U(x)$ by any monotone increasing transformation of U.

Remark 3: If we define the marginal rate of transformation between x^i and x^j MRS_{ij}, $\frac{dx^i}{dx^j}$ along a given indifference curve $U(x) = \text{constant}$, then $MRS_{ij} = -\frac{U_i}{U_j}$ and it follows from lemma 1 that: At the optimum, none of $p_i, p_j, x_i, x_j, U_i, U_j$ is zero then a) the budget constraint is effective and b) $MRS_{i,j} = \frac{P_i}{P_j}$.

5. A PROBLEM IN WELFARE ECONOMICS

In this section we study the market implications of Pareto optimality. We consider an economy with n commodities, m production processes and R consumers.

We shall retain the definitions of consumption and consumers of the previous section. Let x^r denote the vector of n commodities consumed by consumer r and let x denote the matrix whose rows are x^r, $r = 1, ..., R$. Let U^r denote consumer r's utility function. We shall assume that *U^r depends only on x^2*. Let m^r denote the income of consumer r. Let ξ^r be consumers's initial enowment of commodities. We assume that $\xi^r \neq 0$ for all r. We assume that m^r is derived from shares in the profits of production processes and from the sale of the initial endowment.

Let $v = (v^1, ..., v^m)$ denote the levels at which production processes are operated. Let $g^{ij}(v^j)$ denote the net outcome, in terms of good i, when process j is operated at level v^j. We follow the convention of having $g^{ij} \leq 0$ when good i is a net input to process j at level V^j, and of having $g^{ij} > 0$ when good i is a net output of process j at level v^j. The profits of process k, if the price vector $p = (p^1, ..., p^n)$ prevails, is defined to be $\pi^j = \sum_i p^i g^{ij}$. Consumer r's income is given by:

$$m^r = \sum_j B^{rj}\pi^j + \sum_i p^i\xi^{ir},$$

where $\beta^{rj} \geq 0$, $\sum_r \beta^{rj} = 1$ are the shares of the consumers in the profits of processes.

Definition: *Feasible allocation:* An allocation (x,v) is said to be feasible if and only if:

(i) $x^{ri} \geq 0$, $v^j \geq 0$,

(ii) $h^i(x,v) = \sum_r \xi^{ir} + \sum_j g^{ij} - \sum_r x^{ir} \geq 0.$

The definition of feasibility expresses the requirement that no more, of any commodity, is used than what is available and that consumption and process operation levels are non-negative.

Definition: *Pareto Optimum:* A feasible allocation $(\hat{x},\hat{v})$ is said to be Pareto Optimal if there does not exists any other allocations (x',v') such that $U^r(x'^r) \geq U^r(\hat{x}^r)$ for all r with strict inequality for at least one r.

Definition: *Local Competitive Equilibrium:* A non-negative price vector p together with an allocation $(\hat{x},\hat{v})$ is said to be competitive equilibrium if:

Each consumer locally maximizes his utility subject to his budget constraint, i.e. there exists a neighborhood $N^r(\hat{x}^r)$ such that:

(i) $U^r(\hat{x}^r) \geq U^r(x^r)$ for all x^r with $px^r \leq m^r$, $x_i^r \geq 0$, and $x^r \in N^r(\hat{x}^r)$.

(ii) The profits from each process are maximized, i.e.

$$pg^{\cdot\cdot k}(\hat{v}) \geq pg^{\cdot\cdot k}(v) \quad \text{for} \quad v \geq 0$$

(iii) Demand does not exceed supply, if supply exceeds demand for a given commodity then its price is zero, i.e.

(iii.a) $h(\hat{x},\hat{v}) \geq 0$ and

(iii.b) $ph(\hat{x},\hat{v}) = 0.$

We now show that, under certain conditions, a Pareto optimum is attainable by way of competitive equilibrium.

Proposition: *If* 1) U *and* g *are of class* C^2. 2) $(\hat{x},\hat{v})$ *is a Pareto optimum.* 3) *For each consumer* r *there exists a commodity* i *such that* $\hat{U}^r_{x_i} > 0$. 4) $\eta^r \hat{U}^r_{x^r x^r} \eta^{*r} < 0$ *for* $\eta^r = (\eta^{1r}, \ldots, \eta^{\eta r}) \neq 0$. 5) g *are concave. Then there exists a price vector* p *and a system of shares* β^{rj} *such that* $(\hat{x},\hat{v})$ *is a local competitive equilibrium at* p.

Proof: The proof consists of essentially verifying that theorem 4 of this chapter applies.

Conditions 1) and 3) of theorem 4 are satisfied, in view of conditions 1) and 2) of our proposition. Condition 4) of the theorem is easy to verify. To verify condition 2) of the theorem it suffices to note, since U^r depends only on x^r, that condition 3 of the proposition together with the independence of g^{ik} guarantees that the matrix in definition 2) of regularity has maximum rank. Thus there exist $\alpha^r > 0$ and $\hat{\mu} \geq 0$ such that:

(1) $\hat{L}_{x^{ir}} \leq 0$, $\hat{x}^{ir}\hat{L}_{x^{ir}} = 0$, where $L = \sum_r \alpha^r U^r + \sum \hat{\mu}^i h^i(\hat{x},\hat{v})$.

(2) $\hat{L}_{v^j} \leq 0$, $\hat{v}^j L_{\hat{v}^j} = 0$

(3) $\hat{\mu}^i h^i(\hat{x}) = 0$.

By conditions 4) and 5) of the proposition, the function L is locally concave i.e., there exists a neighborhood of (x,v), S(s,v), such that L is concave there. By (1) and (2), using the first order sufficiency theorem, L is locally maximized at (x,v) i.e.

(4) $L(\hat{x},\hat{v}) \geq L(x,v)$ for all $x \geq 0$, $v \geq 0$, $(x,v) \in S$.

Now take $p_i = \dfrac{1}{\sum_i \hat{\mu}^i} \hat{\mu}^i$ and choose $\hat{\beta}^{rj}$ and $\hat{\xi}^r$ such that

$\sum_i p^i \hat{x}^{ir} = \hat{m}^r = \sum_j \beta^{rj}\hat{\pi}^j + \sum_i p^i \hat{\xi}^{ir}$, where $\hat{\beta}$ and $\hat{\xi}$ are redistributions of profit shares and endowments , where $\hat{\pi}^j = \sum_i pg^{ij}(\hat{v}^j)$. Clearly, $p^i \geq 0$, and the choice of $\hat{\beta}^{rj}$ and $\hat{\xi}^{ir}$ is possible (by (3)).

Relation (4) holds, in particular, if we set, on the right hand side, all x's and v's at the optimum exempt for x^{r_0} where r^0 is arbitrary. Thus we have:

$$\alpha^{r_0}U^{r_0}(\hat{x}^{r_0}) - \sum_i \hat{\mu}^i\hat{x}^{ir_0} \geq \alpha^{r_0}U^{r_0}(\hat{x}^{r_0}) - \sum_i \hat{\mu}^i x^{ir_0}, \text{ i.e.}$$

(5) $\alpha^{r_0}(U^{r_0}(\hat{x}^{r_0}) - U^{r_0}(x)) \geq$

$$\geq (\sum_i \hat{\mu}_i)(\sum_i p^i\hat{x}^{ir_0} - \sum p^i x^{ir_0}) = (\sum_i \hat{\mu}_i)(m^r - \sum p^i x^{ir_0}).$$

By (5) $\hat{x}^{r_0}$ locally maximizes $U^{r_0}(x^{r_0})$ subject to $px^{r_0} \leq mx^{r_0}$.
Since r_0 is arbitrary, we have shown (i) in the definition of competitive equilibrium. Noting that (4) holds, in particular, when we set, on the right hand side, all the x's at the optimum and all the v's at the optimum except for an arbitrary j_0. Then we get:

$$(\sum_i \hat{\mu}_i)\sum_i p^i g^{ij_0}(\hat{v}^{j_0}) \geq (\sum_i \hat{\mu}_i)(\sum_i p^i g^{ij_0}(v^{j_0})), \text{ i.e., } \hat{\pi}^{j_0} \geq \pi^{j_0},\ v^{j_0} \geq 0.$$

Note that the concavity of g guarantees that the maximum is global. Thus we have verified condition (ii) in the definition of the competitive equilibrium. Condition (iii.a) follows from feasibility, and condition (iii.b) follows from (3) above. This completes our proof.

□

6. ODDS AND ENDS

This section contains some ideas which I fully intended to develop and which I used to entertain friends in class and during informal discussions. I will review these ideas here in the spirit of putting them up for adoption with the hope that they find suitable homes, perhaps even with me yet.

6.1 SUPER-LAGRANGIANS AND NONLINEAR PRICING

The Kuhn-Tucker theory, Kuhn-Tucker (1951), shows that under concavity a constrained maximization problem implies the maximization of a functional which is linear in the maximand and constraint vectors. The linearity of the functional manifests itself as parametric prices, or linear pricing rules, in economic applications. The separability of decision variables, stemming from independence assumptions on economic behavior, leads to decentralization. In the absence of concavity one may go local in the sense of the proposition at the end of the preceding section or one may augment the Lagrangian or transform the constraints enough to get concavity. This concept appears first in Arrow and Hurwicz (1960). A different direction, changing the form of the Lagrangian, where the term "augmentability" is used is due to Hestenes, see Hestenes (1975) for a full discussion.

It may be useful to illustrate the discussion by using geometric terms. In the case of concavity, it is possible to separate the constraint set and upper contour set above the solution point by using hyperplane i.e. it is possible to separate linearly. In the absence of concavity one may be able to separate nonlinearly. Another way of saying the same thing, is that one may be able to change the variables, separate linearly and transform back. The general question may be stated as follows: let $\hat{x}$ solve a constrained maximization problem. Does there exist a composition of the maximand and constraints which is maximized at $\hat{x}$ subject to no constraints? The answer is yes in the case of concavity. There exist examples of affirmative answers in the absence of concavity, e.g. Arrow-Hurwicz (1960) and Hestenes (1975). What is the boundary of the class of problems with augmentable Lagrangians which we propose to call: *super-Lagrangians*? Under what conditions will the super-Lagrangian be non-separable? The economic meaning of this last question is: under what conditions will a societally efficient allocation dictate a merger of all agents?

6.2. NEXT BEST AND NEAR EQUILIBRIA

In this section we explore an alternative to augmentation. Basically it is a case of: "if you can not be where you wish to be, how much can you change your wishes." Better yet, if we are talking about an actual economy then our statements about preferences and technology are approximate. Are we still within the margin of error if we change the parameterization of the economy so that the economy is at a desirable state.

Suppose that "marginal" conditions are satisfied at a state of economy. Suppose also that some sectors or firms exhibit increasing returns to scale. It is possible then that the state is not a societal optimum (Pareto optimum), in fact such an optimum may not exist for this economy as shown by Brown-Heal (1979).

Brown-Heal (1979) provide a set of sufficient conditions for an "equivalent" economy, to the above economy, to have an optimum. Arrow-Hurwicz (1957) indicate a solution to the problem consisting of concavifying the Lagrangian of the associated maximization problem. So one either perturbs the economy or disturbs the market structure (making prices non-parametric) in order to deal with the "economic rent" accruing from increasing returns.

Here we contrast both approaches locally, which makes it possible to see the nature of the remedy and get by with less assumptions that would have been required had we gone "global". The economy is not perturbed by much and we obtain a local equilibrium which is near the point where marginal conditions hold. The prices are still parametric but a tax-subsidy scheme is imposed. The new state is Pareto Optimal. This uses Smale (1974)results. In a sense, much of what is said here elaborates Brown-Heal (1979) and Smale (1974).

When we stay with the point where the marginal conditions are satisfied but seek a different market structure we use the method of augmented Lagrangians established by Hestenes (see e.g. Hestenes (1975)). We obtain an equilibrium under quantity dependent prices. However, there is no guarantee that it is optimal.

Let R^n be the commodity space. There are m consumers and ℓ firms. A state of economy is a point in $R^{(\ell+m)n}$; an assignment of an n-vector to each agent. Consumer i's preferences are represented by a twice differentiable utility function $U^i(c_i)$, where c_i is consumer i's consumption, i = 1, ..., m. Firm j's technology is given by:

$$f^j(x_j) \leq 0$$

where x_j is the firm's netput vector (a negative component indicates a net input and a positive one indicates a net output); j = 1, ..., ℓ.

A state $Z = (c_1, ..., c_m, x_1, ..., x_\ell)$ is *feasible* if

1) $\omega + \sum_{j=1}^{\ell} x_j - \sum_{i=1}^{m} c_i \geq 0$

2) $c_i \geq 0, f^j(x_j) \leq 0,$

for i = 1, ..., m, j = 1, ..., ℓ, where ω is the initial endowment vector.

A feasible state is said to be *Pareto Optimal*, *Optimal* for short, if there does not exist a feasible state superior to it. Departing from Smale's (1974) terminology we define a *classical equilibrium* a price vector $\hat{\rho} \in R^n_+$, with $\sum \rho_i^2 = 1$, and a state Z of the economy such that the "market conditions" are satisfied, i.e.

3) $\hat{U}^i_c = \lambda_i \hat{\rho}$

4) $\mu_j \hat{f}^j_x = \hat{\rho}$

5) $\omega + \sum_j \hat{x}_j = \sum_i \hat{c}_i$

6) $\hat{c}_{is} > 0, \qquad f^j(\hat{x}^j) = 0, \quad \hat{x}^j_s \neq 0, \qquad s = 1, ..., n.$

where $\hat{U}^i_c = \frac{\partial}{\partial c_i} U^i \big|_{c_i = \hat{c}_i}$, $\hat{f}^j_x = \frac{\partial}{\partial x_j} f^j \big|_{x_j = \hat{x}_j}$ and where $\lambda_i > 0, \mu_j > 0,$

i = 1, ..., m, j = 1, ..., ℓ. An *equilibrium* is defined as a price state pair ρ, Z such that consumers maximize utility subject to budget constraint, producers maximize profits subject to technological feasibility ($f^j \leq 0$) and where inequalities 1) and 2) hold. An *interior local equilibrium* is defined to be an equilibrium with 5) and 6) replacing 1) and 2) and where the maximization is local.

Clearly, assuming the appropriate constraint qualification, an optimum is a classical equilibrium. It is also true that an interior local equilibrium is a classical equilibrium. Let us recall the definition of a *stable critical point* of a smooth function to be a critical point of the function where the Hessian is negative semi definite. Assuming local quasi concavity of utility functions and local quasi convexity of the transformation functions, f^j, is more than enough to guarantee that classical equilibria are stable. The class of economies where a stable classical equilibrium is not an equilibrium is not big, Smale (1974). Except for these economies, a classical equilibrium is a local optimum.

If $(\hat{Z}, \hat{\rho})$ is a classical equilibrium then $\hat{Z}$ is a critical point of the function:

7) $L(Z; \rho, \alpha, \mu) = \sum \alpha_i u^i + \rho \cdot (\omega + \sum_j x_j - \sum_j c_i) - \sum_j \mu_j f^j$, where $\alpha_i = 1/\lambda_i$.

Assume that the Hessian of L with respect to Z, at $\hat{Z}$, $\hat{L}_{ZZ}$, is negative semi definite then there exists a linear function S(Z) such that $\mathcal{L} = L + S(Z)$ has a critical point at $\mathcal{Z}$ where the Hessian of L at $\mathcal{Z}$ is negative definite (see Gibson (1979) and Golubitsky and Guillemin (1973). Furthermore $\mathcal{Z}$ is a feasible state and is close to $\mathcal{Z}$. The function $\mathcal{L}$ attains a local maximum at $\mathcal{Z}$. Let us write $S(Z) = \delta c + \sigma x$. Then each consumer maximizes u^i subject to $(\rho - \delta)c_i \leq m$. Each producer maximizes $(\rho + \delta)x_j$ subject to $f^j(x_j) \leq 0$.

Assuming nonsatiation (local) we can, using the standard argument show that $\mathcal{Z}$ is a Pareto Optimal. Using the steps of proof in Smale (1974) we can show that $\mathcal{Z}$ is close to $\hat{Z}$. We propose to call $(\mathcal{Z}, \hat{\rho}, \delta, \sigma)$ a *near equilibrium* and $\mathcal{Z}$ a *next best*.

We have outlined the proof of:

Proposition. *Suppose the "nonconvexities" in the economies are exactly balanced by the "convexities". Assume nonsatiation (local) and enough independence to guarantee an appropriate "constraint qualification". Suppose the state price pair* $(\hat{Z}, \hat{\rho})$ *is a point of classical equilibrium. Then there exists a state* $\mathcal{Z}$ *close to* $\hat{Z}$ *and subsidy-tax scheme* (σ, δ) *such that consumers pay the producers to produce and sell at marginal cost prices. So that* $\mathcal{Z}$, *consumer price* $\hat{\rho} - \delta$ *and producer price* $\hat{\rho} + \sigma$ *define a local competitive equilibrium. Furthermore,* $\mathcal{Z}$ *is a local Pareto optimum.*

Remarks.

1) The balancing of convexities and nonconvexities means here that the Hessian of the Lagrangian (e.g. (7) above) is singular.

2) The tax subsidy scheme is very similar to the usual proposal in marginal cost pricing literature.

3) The new, perturbed, Lagrangian is that of an "equivalent" economy in the sense of Brown-Heal (1979).

4) Associating nonconvexities in production with increasing returns one sees a "rent" which accrues to consumers who share it, some, with poor monopolists.

5) Now suppose that instead of perturbing the Lagrangian we augment it, so we consider

$$8)\quad \tilde{\mathcal{L}} = \sum \alpha_i u^i + \hat{\rho} \cdot (\omega + \sum_j x_j - \sum_i c_i) - \sum_j \mu_j f^j + \frac{\partial}{2} (\sum_{i=1}^{n} (\omega_i + \sum_j x_j - \sum_i c_i)^2 - \sum_j (f^j)^2).$$

We are still assuming that $\hat{\mathcal{Z}}, \hat{\rho}$ is a classical equilibrium. Hestenes (1975) shows that a value of δ, say δ^0, exists so that $\tilde{\mathcal{L}}$ is locally maximized at $\hat{\mathcal{Z}}$. It then can be seen that the prices of the form:

$$9)\quad \tilde{\rho}_i = \hat{\rho}_i + \frac{\delta}{2} (\omega_1 - \sum_j x^*_j - \sum c^*_i)$$

with (*) denoting the putting every "other" agent at their optimum, we get that $\hat{Z}$ is a Nash equilibrium in a nonparametric price structure. A global version of augmentation was first introduced in Arrow-Hurwicz (1957).

6) Thus, in view of the proposition and remark 4) we either collect the rent and redistribute it or allow monopolistic elements to appropriate it.

6.3. WHAT ADAM SMITH REALLY SAID

The above two sections are in the spirit of that part of Adam Smith's Wealth of Nations which no one seems to read. The Wealth of Nations issues from a spirit of cooperation in the presence of externalities. What is obvious, in fact nearly intellectually devoid of content, is that in the absence of any interdependence what is good for the group is good for the individual. What is more challenging is to prove the following:

To the extent that there interdependence, a cooperative structure is called for.

6.4. WHAT MARX SHOULD HAVE SAID

In several works, notably the excellent treatments by Roemer (1980), (1982), and (1986) and by Abraham-Frois and Berrebi (1976), attempts were made to solve Marx's riddle about the existence of exploitation under an institutionally noncoercive economic regime. This is done, more or less, by reconciling the dual theories of value that Marx believed in.

In this section we suggest a simple explanation of exploitation which we say to occur when real wages are less than the marginal product of labor at equilibrium in the absence of institutional coercion. The idea is to simply recognize the subsistence constraint on workers' behavior.

Assume that a worker determines the supply of his labor in the classical fashion of maximizing his utility function, with income and leisure as arguments, subject to a subsistence constraint. let $u(m, s)$ denote that utility function, where m is income and where s is leisure. The problem is to maximize u subject to the constraints: $m = w\ell$, $\ell + s = L$, $\ell \geq 0$, $s \geq 0$, and $m \leq m^0$. here w is the real wage level, ℓ is the work, L is the length of the time period under study and m^0 is the subsistence real income. One might think of m^0 as physiologically determined or psychologically determined. The latter makes it safer to state that the subsistence constraint is in fact binding, albeit at differing levels, for a majority of workers in modern times.

The problem may be restated as: maximize $u(w\ell, L - \ell) = v(\ell)$, subject to:

$$w\ell < m^0 \text{ and } 0 \leq \ell \leq L.$$

It the subsistence constraint is ineffective then we obtain the usual labor supply curve which is positively sloped up to a certain wage level say w^* where it becomes negatively sloped. Let w^0 be a real wage level where the subsistence constraint becomes effective.

Then the best value of ℓ is the only feasible value for ℓ; $\ell^0 = m^0/w^0$. If w is below w^0 then the subsistence constraint will remain effective and the supply of labor is given by: $\ell = m^0/w$.

Finally, let $w\dagger$ be the real wage level where one works all day and night in order to survive. Clearly, $\ell = 0$ if $w < w\dagger$ since the worker has just expired.

Thus, assuming enough smoothness and convexity, the supply curve of labor is given by

$$\ell(w) = \begin{cases} h(w), & w^0 \leq w \\ m^0/w, & w\dagger \leq w \leq w^0 \\ 0, & w < w\dagger \end{cases}$$

The function h(w) is the result of the interior maximization problem stated above.

Now consider a producer who utilizes labor and other inputs to produce a single output. If the producer maximizes profits then, under the usual assumptions, his demand for labor is given by a downward sloping marginal productivity curve for labor which we denote by $g(w, \alpha)$, where the vector α stands for output, other inputs and their real prices.

The point $(w\dagger, L)$ represents the "classical" iron wage equilibrium since the adjustment process stops at that point. Another plausible situation is depicted in the diagram below.

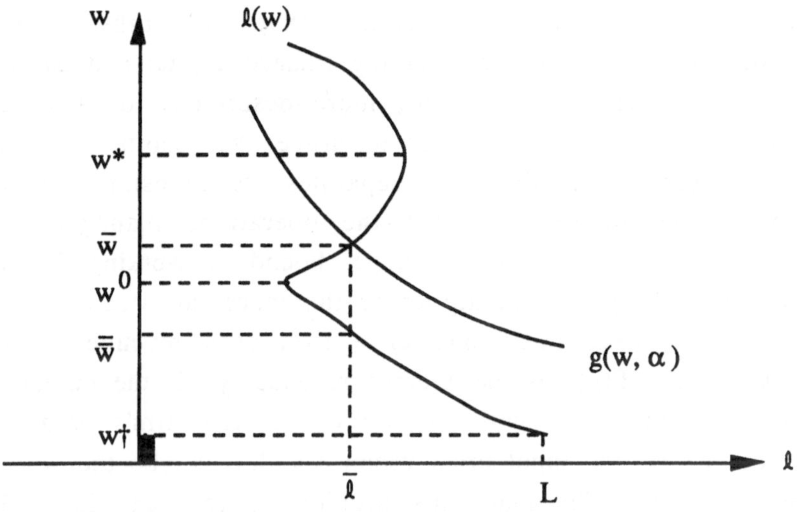

Figure 1. Supply curve of labor with subsistence constraint.

The equilibrium quantity of labor is $\bar{\ell}$. But, with perfect information, the employer does not have to pay $\bar{w}$ and ends up paying $\bar{\bar{w}}$.

Thus exploitation is plausible in the absence of coercion by the employer. It is possible because the worker must sell his services to survive.

Another way to show the plausibility of exploitation is to assume a constant returns to scale homogeneous production function. Start with an equilibrium at w^0. Suppose prices go up so that $w < w^0$. The demand for and the supply of labor increases. Labor share is constant since $w\ell = m^0$. But real output increases even if other factors employed remain fixed, assuming a positive marginal productivity of labor. The shares of other factor increase since the output must be exhausted. Thus exploitation, in the classical sense, occurs even if factors other than labor are productive.

6.5. AND BOHM-BAWERK TOO

Much of the discussion on roundaboutness has to do with the productivity of capital, see Bohm-Bawerk (1890) and thus confound Marx. If one thinks of capital as working finance capital and adopts a theory of the firm which is, perhaps, more descriptive of a modern firms behavior then it will be easier to argue that capital is indeed productive. Assume that the firm separates its investment budget from its operations budget and that the operations manager is to maximize his revenue subject to and upper bound of working financial capital. Then the Lagrange multiplier is the marginal productivity of finance capital. To see this, consider a firm with revenue function $G(x) = p \cdot y$, where $P(y)$ is the average revenue, y is the output and x is the n input vector with $y = f(x)$ as the firm's production function. Let q be the input price vector and let k be the working capital upper bound. The manager's problem is to maximize $G(x)$ subject to $k - q \cdot x \geq 0$. Assume that G is C^2 and that $q > 0$. Then, at a solution point $x > 0$ to the maximization problem there exists a unique λ such that: $G_i = \lambda q_i$, $i = 1, ..., n$. Assume that $G_i > 0$ then $\lambda > 0$ and $k = q \cdot x$. If we change m then x will change smoothly and so will λ. Then $dG = G_x \cdot dx$ and $dk = q \cdot dx$ if q is fixed. By the first order conditions: $dG = \lambda q \cdot dx = \lambda dk$. Thus $\partial G/\partial k = \lambda$ and λ is the marginal revenue product of working capital . If the firm were competitive then we have $\lambda = p \dfrac{\partial f}{\partial k}$. If we choose a model where the firm maximizes profits subject to an upper bound on working capital then under the same conditions stated above we have: $G_i = (1 + \mu)q_i$ and $d\pi = \sum_i G_i dx_i - q_i dx_i) = (1 + \mu)q \cdot dx - q \cdot dx = \mu q \cdot dx$, where π is the profit function and where μ is the Lagrange multiplier for the new problem. But $dk = q \cdot dx$ when q is fixed. Thus $\frac{\partial \pi}{\partial k} = \mu$. The function μ is the marginal profitability of finance capital.

7. MORE ABOUT CONCAVITY AND ITS GENERALIZATIONS

We have seen, assuming as we do that functions are C^2, that if the maximand and constraints are concave, then a solution to the inequality constrained maximization problem (ICM) is a solution to the problem of maximizing the Lagrangian (ML) associated with that ICM. We discuss here some attempts at proving that a solution of ICM is a solution of ML under conditions that is weaker than concavity.

It has been shown, see Fenchel (1953), Arrow-Enthoven (1961), and Luenberger (1968) that quasi-concavity of the maximand and constraints would not accomplish our objective. This is not just because a non-negative linear combination of quasi-concave functions is not necessarily quasi-concave. One needs also non-negative gradints in order to show that. For even a quasi-concave lagrangian need not be maximized at a solution of ICM. Indeed, consider the problem of maximizing $[(1 - x)^2 + 4(x + y)]^{1/2}$ subject to $x - 1 \geq 0$. The Lagrangian in this case is quasi-concave and it is not maximized at the point that solves ICM. But, as Fenchel (1953) showed and as Arrow-Enthoven (1961) noted, this Lagrangian cannot be transformed into a concave function by way of a monotone increasing transformation i.e. in the words of Kannai (1977) it is not concavifiable.

So then is it true that if a solution to ICM is a solution to ML then the Lagrangian is concavifiable? Not quite, but nearly so. To answer the question, let $g: R^n \to R$ be a C^2 function which is quasi-concave. Let $F(x) = f \circ g$ be a monotone strictly increasing C^2 function. Using Fenchel, (1953) section 8 pp. 127-137, we can easily show that if F is concave then:

(A) If x^* is a critical point of g then $x^* = \text{argu}\max(g)$.

(B) $\sum_{i,j} g_{ij}(x)a_i a_j < 0$ for a satisfying $\sum_i g_i(x)a_i = 0$,

where x is any point in the domain of g. Further, the rank of the unconstrained Hessian of g is at most r where $r - 1$ is the rank of the constrained Hessian.

(C) $\left(\frac{f'(c)}{f''(c)}\right) \leq \inf_{g(x)=c} \left(\frac{S_r}{k^2 S^*_{r-1}}\right)$,

where $k^2 = \sum_i g_i^2$, S_r is the sum of all principal minors of (g_{ij}) of order r, and where S^*_{r-1} is the sum of the principal minors of (g_{ij}) constrained by $\sum g_i a_i = u$.

Conversely if A, B, *and* C *hold then* f *is concave, i.e.* g *is concavifiable.*

An answer to our original question can now be attempted. Suppose 1) every solution to ICM is a solution to ML, 2) Some form of constraint qualification holds, 3) Conditions B and C of Fenchel, stated above, are satisfied. Then the Lagrangian is concavifiable.

If a C^2 function satisfies the first part of condition B above then it is quasi-concave. The converse is, clearly, not true. At a point where the constrained Hessian of a function f is zero for some nonzero "increment", we say that the Hessian is *singular*. Otherwise, if it is never zero for nonzero "increment", we say that the function is *Samuelson-regular*. For Samuelson-regular functions, the first part of condition B implies that the function is quasi-concave. Samuelson, see Samuelson 91947), defines a *regular maximum* at a point as a critical point where the Hessian is negative definite. In honor of this, we term a function that satisfies condition B a *Samuelson-quasi-concave* function. Clearly a Samuelson quasi-concave function is quasi-concave. It is also true that all maxima of Samuelson-quasi-concave functions are Samuelson-regular.

8. NOTES ON THE LITERATURE

The method of proof of the scalerization lemma for vector maxima was suggested to the author by Professor Leonid Hurwicz in 1963. The second order condition for vector maxima was circulated as a preprint, El Hodiri (1966) and was included in the first version of these notes, El Hodiri (1971). Unfortunately it was, I am sure independently, rediscovered and republished by others. On the other hand an original thinker may be forgiven a lapse in scholarship.

PART II

VARIATIONAL PROBLEMS

CHAPTER 5

THE PROBLEM OF BOLZA WITH EQUALITY CONSTRAINTS

We shall state, here, some theorems that characterize solutions to the problem of Bolza in the calculus of variations. The proofs of some of these theorems will only be briefly outlined. The reader may refer to Bliss (1930), (1938), and (1946) and to Pars (1962) for a more detailed presentation. By way of introduction we discuss an unconstrained problem in the calculus variation, in section 1. In section 2, we state the problem of Bolza. In section 3 we discuss first order necessary conditions and in section 4 we state the necessary conditions of Weierstrass, Clebsch, and Mayer. In section 5, we state second order sufficient conditions. In section 6, we characterize solutions to problem A'.

1. UNCONSTRAINED PROBLEM

In this section we study characterization of constrained extrema of an integral. The results indicate the type of results expected for the constrained case and are utilized in the derivation of these results.

Let $T = [a,b]$ be a subset of the real line. Consider the set of functions $y: T \to E^n$ and the set of functions $z: T \to E^n$. Let the function $f(t,y,z): E^{2n+1} \to E^1$ be defined for all values taken by $t \in T$, $y \in E^n$ and $z \in E^n$. Let A and B be two points in E^n. Define the set of admissible arcs M as follows:

$M = \{y$: 1) y is sectionally smooth, 2) the integral $\int_a^b f(t,y,\dot{y})\, dt$

has a finite value, 3) $y(a) = A$ and $y(b) = B\}$, where $\dot{y} = dy/dt$.

The problem is to characterize the extrema of:

$$J[y] = \int_a^b f(t,y,\dot{y})\,dt.$$

From now on, we shall confine our attention to maxima of J[y]. We distinguish two main types of maxima: global and local.

Definition 1: An arc $\bar{y} \in M$ is said to be a global maximum of J[y] if $J[\bar{y}] \geq J[y]$, for all $y \in M$.

Definition 2: An arc $\bar{y}$ is said to be a local maximum of J[y] if $J[\bar{y}] \geq J[y]$, for all $y \in M'$, where $M' \subset M$, $y \in M'$.

We distinguish two types of local maxima: strong and weak. The distinction is based on the definition of M'.

Definition 3: We define a strong ε-neighborhood of $\bar{y}$, $N_0(\bar{y},\varepsilon)$ as follows: $N_0(\bar{y},\varepsilon) = \{y \in M: d(y(t),\bar{y}(t)) \leq \varepsilon, (t \in R)\}$ where $d(y(\bar{t}),\bar{y}(\bar{t}))$ is the Euclidian distance, i.e. equals $\sqrt{\sum_i (y_i(\bar{t}) - \bar{y}_i(\bar{t})^2)}$, $\varepsilon > 0$.

Definition 4: A weak ε-neighborhood of $\bar{y}$, $N_1(\bar{y},\varepsilon)$, is defined as follows:

$$N_1(\bar{y},\varepsilon) = \{y \in N_0(\bar{y},\varepsilon): d(\dot{y}(t),\dot{\bar{y}}(t)) \leq \varepsilon, (t \in T)\}.$$

We are now ready to define the two types of local maxima.

Definition 5: An arc $\bar{y}$ is said to be a weak local maximum of J[y] if $J[\bar{y}] \geq J[y]$, $(y \in N_1,(\bar{y},\varepsilon))$ for some $\varepsilon > 0$.

Definition 6: An arc $\bar{y}$ is said to be a strong local maximum of J[y] if $J[\bar{y}] \geq J[y]$, $(y \in N_0(\bar{y},\varepsilon))$ for some $\varepsilon > 0$.

<u>Remark 1)</u> If $\bar{y}$ is a global maximum then it is a strong local maximum, and if $\bar{y}$ is a strong local maximum, then it is a weak local maximum. This follows from the fact that $N_1(\bar{y},\varepsilon) \subset N_0(\bar{y},\varepsilon) \subset M$.

1.1 FIRST ORDER NECESSARY CONDITIONS

In this section we derive a necessary condition for a weak local maximum in terms of first derivatives of J. The condition, by remark 1 above, is also necessary for strong local maxima and global maxima.

Theorem 1. *If* f *is continuously differentiable and if* $\bar{y}$ *is a weak local maximum of* J[y], *then there exist constants* C_i *such that:*

(0)

$$\bar{f}_{\dot{y}_i} - \int_a^t \bar{f}_{y_i} = C_i, \quad i = 1, \ldots, n, \text{ where}$$

$$\bar{f}_{y_i} = \left.\frac{\partial f}{\partial y_i}\right|_{y = \bar{y},\, \dot{y} = \dot{\bar{y}}} \quad \text{and} \quad \bar{f}_{\dot{y}_i} = \left.\frac{\partial f}{\partial \dot{y}_i}\right|_{y = \bar{y},\, \dot{y} = \dot{\bar{y}}}.$$

Proof: Let $\eta(t)\colon T \to E^n$ be a vector valued function, whose components are $\eta_i(t)$, such that $\eta(a) = \eta(b) = 0$ and such that $\eta_i(t)$ are sectionally smooth. For $y \in N_1$, we may write, $J[y] = \phi(\alpha)$. Note that $\phi(\alpha)$ is differentiable and that $\phi(0) = J[\bar{y}]$ is a maximum value of ϕ. Thus $\phi'(0) = 0$. The rest of the proof consists of computing $\phi'(0)$. Differentiating under the integral sign we get:

$$\phi'(\alpha) = \int_a^b \Big(\sum_i f_{y_i}\eta_i + \sum_i f_{\dot{y}_i}\dot{\eta}\Big)\, dt,$$

Thus

(1) $\phi'(0) = \int_a^b (\sum_i \bar{f}_{y_i}\eta_i + \sum_i f_{\dot{y}_i}\dot{\eta}) \, dt = 0.$

Integrating $\int_a^b f_{y_i}\eta_i \, dt$ by parts we get:

(2) $\int_a^b \bar{f}_{y_i}\eta_i = \eta_i \int_a^t \bar{f}_{y_i} \, dt \Big|_a^b - \int_a^b (\dot{\eta}_i \int_a^t \bar{f}_{y_i} dt) \, dt.$

But $\eta_i(a) = \eta_i(b) = 0$, and the first term of (2) vanishes. Thus we have:

(3) $\int_a^b \bar{f}_{y_i}\bar{\eta}_i \, dt = - \int_a^b (\dot{\eta}_i \int_a^t \bar{f}_{y_i} dt) \, dt.$

Substituting, from (3), into (1) we have:

(4) $\int_a^b \sum_i (\bar{f}_{\dot{y}_i} - \int_a^t \bar{f}_{y_i} dt) \, \dot{\eta}_i dt = 0.$

Define the functions $M_i(t)$ and the constants C_i as follows:

(5.1) $M_i(t) = \bar{f}_{\dot{y}_i} - \int_a^t \bar{f}_{y_i} dt,$

(5.2) $C_i = \frac{1}{b - a} \int_a^b M_i(t) \, dt.$

We shall show that C_i are the constants whose existence is asserted by the theorem, i.e. they are such that $M_i = C_i$.

Since $\eta(t)$ is arbitrary, we may pick

(6) $\eta_i(t) = \int_a^b (M_i - C_i) \, dt$

Note that such a choice of η_i is appropriate, since η_i are sectionally smooth and since:

(7.1) $\eta_i(a) = \int_a^a (M_i - C_i)\, dt = 0$, and

(7.2) $\eta_i(b) = \int_a^b (M_i - C_i)\, dt = \int_a^b M_i dt - \int_a^b C_i dt$

$$= \int_a^b M_i dt - (b - a)C_i$$

$$= \int_a^b M_i dt - (b - a)\frac{1}{b - a}\int_a^b M_i dt = 0$$

Now, $\dot{\eta}_i(t) = \frac{d}{dt}\int_a^t (M_i - C_i)dt = M_i - C_i$. Thus (4) becomes

(8) $\int_a^b \sum_i M_i(M_i - C_i)\, dt = 0$

But (8) is equivalent to

(9) $\int_a^b \sum_i (M_i - C_i)^2\, dt = 0$

The equivalence of (8) and (9) is due to:

$$\int_a^b \sum_i (M_i - C_i)^2\, dt = \int_a^b \sum_i M_i(M_i - C_i)\, dt - C_i \int_a^b \sum_i (M_i - C_i)\, dt =$$

$$= \int_a^b \sum_i M_i(M_i - C_i) - C_i\eta(b) = \int_a^b \sum_i M_i(M_i - C_i)\, dt,$$

(by 7.2). But the integrand and interval of integration in (9) are non-negative, thus:

(10) $\sum_i (M_i - C_i)^2 = 0.$

From (10) it follows that $M_i = C_i$. □

<u>Remark</u>: It follows from the theorem of this section that if $\bar{y}$ is a maximizing arc then

(11) $\bar{f}_{y_i} - \frac{d}{dt}\bar{f}_{\dot{y}_i} = 0.$

The equations (10) and (11) are the Euler equations.

1.2. FIRST ORDER SUFFICIENT CONDITIONS

In this section we show that $\bar{y}$ is a stationary arc for $J[\bar{y}]$, i.e. if it satisfies the Euler equations and if f is concave in y and $\dot{y}$ then it furnishes $J[y]$ with a weak relative maximum. Let us recall from Chapter 4 of these notes that if a function $g(x)$ is differentiable and concave, then

(1) $g(x) - g(\bar{x}) \leq \sum_i \bar{g}_i(x_i - \bar{x})$, where $\bar{g}_i = \left.\frac{\partial g}{\partial x_i}\right|_{x = \bar{x}}$.

Theorem 2. *If* f *is continuously differentiable and concave in* y *and* $\dot{y}$, *and if* $\bar{y}$ *is a stationary arc for* J[y] *then* $\bar{y}$ *is a global maximum of* J[y].

Proof: Let y be an admissible arc ($y \in M$), then:

(2) $J[y] - J[\bar{y}] = \int_a^b [f(t,y,\dot{y}) - f(t,\bar{y},\dot{\bar{y}})]\, dt.$

By concavity of f (and differentiability) we have:

(3) $f(t,y,\dot{y}) - f(t,\bar{y},\dot{\bar{y}}) \leq \sum_i \bar{f}_{y_i}(y_i - \bar{y}_i) + \sum_i f_{\dot{y}_i}(\dot{y}_i - \dot{\bar{y}}_i)$, $(t \in [a,b])$.

Integrating both sides of (3) we get (in view of (2)):

$$(4) \quad J[y] - J[\bar{y}] = \int_a^b \left[\sum_i \bar{f}_{y_i}(y_i - \bar{y}_i) + \sum_i f_{\dot{y}_i}(\dot{y}_i - \dot{\bar{y}}_i) \right] dt.$$

Since $\bar{y}$ is stationary, the right hand side may be written as:

$$(5) \quad (a,b,) \left[\sum_i \frac{d}{dt} \bar{f}_{\dot{y}_i}(y_i - \bar{y}_i) + \sum_i f_{\dot{y}_i}(\dot{y}_i - \dot{\bar{y}}_i) \right] dt$$

Integrating the first term by parts, the expression (5) becomes:

$$(6) \quad \sum_i (y_i - \bar{y}_i) f_{\dot{y}_i} \Big|_a^b - \int_a^b \sum_i f_{\dot{y}_i}(\dot{y}_i - \dot{\bar{y}}_i)\, dt + \int_a^b \sum_i f_{\dot{y}_i}(\dot{y}_i - \dot{\bar{y}}_i)\, dt.$$

Thus we have (in view of (4), (5), and (6)):

$$(7) \quad J[y] - J[\bar{y}] \leq \sum_i (y_i - \bar{y}_i) f_{\dot{y}_i} \Big|_a^b .$$

But y is in M, thus $y(a) = \bar{y}(a)$ and $y(b) = \bar{y}(b)$, i.e. the right hand side of (7) is zero. Thus

$$J[y] - J[\bar{y}] \leq 0,$$

and since y is an arbitrary arc in M, $\bar{y}$ is a global arc for $J[y]$. □

1.3. THE NECESSARY CONDITION OF WEIERSTRASS

We define the Weierstrass excess function at maximal arc $\bar{y}$ as the change in the value of the integrand f when $\dot{\bar{y}}$ is replaced by an arbitrary vector p of real numbers:

$$E(t,\bar{y},\dot{\bar{y}},p) = f(t,\bar{y},\dot{\bar{y}}) - f(t,\bar{y},p) - (p - \dot{\bar{y}})\bar{f}_{\dot{y}}.$$

Theorem 3: *If* f *is of class* C^2 *and if* $\bar{y}$ *provides a weak local maximum for* $J[y]$ *in a weak neighborhood* $N_1(\bar{y})$, *then:* $E(t,,\bar{y},\dot{\bar{y}},p) \geq 0$ *for all* p *and for all* $t \in [a,b]$.

Proof: The proof is by contradiction. We shall assume that E is negative for some p_0 at $t = t_0$, $t_0 \in [a,b]$. Then we may find an arc $\tilde{y}(t) \in N_1$ such that $J[\tilde{y}] > J[\bar{y}]$. Define:(1)

$$\tilde{y}(t;h) = \begin{cases} \bar{y}(t), & t \in [t_0, b] \\ (t - t_0)\cdot p_0 + \bar{y}(t_0), & t \in [t_0 - h, t_0] \\ (t - a)\cdot k + \bar{y}(t), & t \in [a, t_0 - h], \end{cases},$$

where h denotes a real number with $t_0 - h > a$ and where k is an n-vector function defined by:

$$(*) \quad \bar{y}(t_0 - h) + (t_0 - a - h)\cdot k(h) = \bar{y}(t_0) - p_0 h.$$

Using the definition of $\tilde{y}(t;h)$ we shall write the difference between $J[\tilde{y}]$ and $J[\bar{y}]$ as a function $\omega(h)$, of a real number h. We then show that $\omega(0) = 0$ and $\omega'(0) > 0$ thus showing that ω is a monotone increasing in a neighborhood of $h = 0$ and that for $h > 0$, in that neighborhood, $\omega(h) > 0$. But then $J[\tilde{y}] - J[\bar{y}] = \omega(h) > 0$, which would complete our proof.

We write $\omega(h) = J[\tilde{y}] - J[\bar{y}]$. Clearly $\omega(0) = 0$, since $\tilde{y}(t,0) = \bar{y}(t)$. But,

$$\omega(h) = \int_a^b f(t,\tilde{y}(t,h), \dot{\tilde{y}}(t,h)) - f(t,\bar{y},\dot{\bar{y}})\, dt = \int_a^{t_0 - h} f(t,\tilde{y},\dot{\tilde{y}}) - f(t,\bar{y},\dot{\bar{y}})\, dt$$

$$+ \int_{t_0 - h}^{t_0} f(t,\tilde{y},\dot{\tilde{y}}) - f(t,\bar{y},\dot{\bar{y}})\, dt + \int_{t_0}^{b} f(t,\tilde{y},\dot{\tilde{y}}) - f(t,\bar{y},\dot{\bar{y}})\, dt$$

The last term is zero, by (1) and $\tilde{y} = \begin{cases} p_0, & t \in (t_0 - h, t_0) \\ \dot{\bar{y}} + k, & t \in (a, t_o - h) \end{cases}$ and

we write:

$$(2) \quad \omega(h) = \int_a^{t_0 - h} f(t, \bar{y} + (t - a)\cdot k, \dot{\bar{y}} + k) + $$

$$+ \int_{t_0 - h}^{t_0} f(t, \bar{y}(t_0) + (t - t_0)\cdot p_0)\, dt - \int_a^{t_0} f(t, \bar{y}, \dot{\bar{y}})\, dt.$$

Differentiating under the integral sign in (2) with respect to h we get:

$$(3) \quad \omega'(h) = \int_a^{t_0 - h} [(t - a)\cdot f_y(t, \tilde{y}, \tilde{y})k_h + f_{\dot{y}} k_h]dt - $$

$$- f(t_0 - h, \bar{y}(t_0 - h) + (t_0 - h - a)\cdot k, \dot{\bar{y}}(t_0 - h) + k) +$$

$$+ f(t_0 - h, \bar{y}(t_0) - h\cdot p_0, p_0) \qquad \text{where} \quad k_h = \frac{d}{dh} k.$$

Evaluating (3) at $h = 0$, we have:

$$(4) \quad \omega'(0) = \int_0^{t_0} (t - a)\cdot(\bar{f}_y + \bar{f}_{\dot{y}})k_h^0 + f(t_0, \bar{y}(t_0), p_0) - f(t_0, \bar{y}(t_0), \dot{\bar{y}}(t_0))$$

where $k_h^0 = k_h |_{h=0}$.

By the maximality of $\bar{y}$ we get, by theorem 1:

$$f_y = \frac{d}{dt} f_{\dot{y}}.$$

Thus we may write: $\int_a^{t_0} [(t - a)\bar{f}_y + \bar{f}_{\dot{y}}]dt$ as $\int_a^{t_0} [(t - a)\cdot(\frac{d}{dt} + f_{\dot{y}})]\, dt +$ $\int_a^{t_0} \bar{f}_{\dot{y}}\, dt$. Integrating the first term by parts we get:

$$\int_a^{t_0} [(t - a)\frac{d}{dt} f_{\dot y}\, dt = (t - a)\cdot\bar f_{\dot y}\Big|_a^{t_0} = (t_0 - a)\cdot\bar f_{\dot y}.$$

So we have:

$$(5)\qquad \int_a^{t_0} [(t - a)\cdot\bar f_y + \bar f_{\dot y}]\, dt = (t - a)\cdot\bar f_{\dot y}\Big|_a^{t_0} = (t_0 - a)\cdot\bar f_{\dot y}.$$

Differentiating both sides of (1) with respect to h we have:

$$- \dot{\bar y}(t_0 - h) - k + (t_0 - a - h)k_h = -p_0.$$

Evaluating at $h = 0$, we get:

$$(t_0 - a)k_h^0 = \dot{\bar y}(t_0) - p_0 + k(0).$$

But evaluating (*) at $h = 0$, we get:

$$y(t_0) + (t_0 - a)\, k(0) = y(t_0),$$

and $k(0) = 0$, since $a - t_0 > 0$. Thus

$$(6)\qquad (t_0 - a)\cdot k_h^0 = \dot{\bar y}(t_0) - p_0.$$

Substituting from (5) and (6) into (4) we have:

$$\begin{aligned} \omega'(0) &= (t_0 - a)\cdot k_h^0 \bar f_{\dot y} + f(t_0,\bar y(t_0),\dot{\bar y}(t_0),p_0) - f(t_{0,0},\bar y(t_0),\dot{\bar y}(t_0)) \\ &= (\dot{\bar y}(t_0) - p_0)\bar f_{\dot y} + f(t_0,\bar y(t_0),p_0) - f(t_0,\bar y(t_0),\dot{\bar y}(t_0)) \\ &= E(t_0,\bar y(t_0),\dot{\bar y}(t_0),p_0) > 0 \end{aligned}$$

Since ω' is continuous, there exists a neighborhood $S(0)$ of $h = 0$ where $\omega'(h) > 0$, choose $h \in S(0)$ such that $\tilde y \in N_1$. □

1.4 LEGENDER NECESSARY CONDITIONS

Theorem 4. *If* f and $\bar{y}$ *are as in theorem 3, then* $\eta'\bar{f}_{\dot{y}\dot{y}}\eta \leq 0$ *for all* η *for* $t \in [a,b]$.

Proof: The proof is by contradiction and it uses the construction of theorem 3. Suppose there exists $t_0 \in [a,b]$ and $0 \neq \eta_0 \in E^n$ such that $\eta_0^* f_{\dot{y}\dot{y}}(t_0,\bar{y}(t_0),\dot{\bar{y}}(t_0))\eta_0$ is positive. Then using Taylor's theorem of the second order we may show that the Weierstrass E function is positive at t_0 for some value of p. We then may construct an arc in N_1 which gives J a larger value than $J[\bar{y}]$, as we did in the proof of theorem 3, which would contradict our hypothesis.

1.5 SUFFICIENT CONDITIONS

In this section we show, under some additional assumptions, that the Weierstrass condition is also sufficient. We also prove a second order sufficient condition. In preparation for this we start by some notations, definitions, and preliminary results.

Definition: We define a *stationary arc* as an arc that satisfies the Euler equations of theorem 3.

Notation: Let S = {the set of stationary arcs}.

Definition: (Fields of a functional): Let $G \subset R^{n+1}$ be the set of values taken by (t,y(t)). Consider D, G, and the function $q(t,y)$: $D \to R^n$. The pair <D,q> is said to be a field of J[y] if and only if:

1) $q \in C^1$ on D,

2) The Hilbert integral, $I_C = \int_C [(f(t,y,q) - q \cdot f_y)] \, dt + f_{\dot{y}}(t,y,q) \cdot dy$,

where $dy = (dy_1, dy_2, ..., dy)$ and where the $a \cdot b$ denotes

$\sum_{i=1}^{n} a_i b_i$, is independent of the path of integration.

Definition: (A set covered by a field, slope): If $<D,q>$ is a field, then we say that D is covered by a field and that q is the slope function of the field.

Definition: Let $J = \{y \mid y = \phi(t,\beta), \beta \in R^n, \dot{y} = q(t,y)\}$. J is called the trajectory of the field.

Remark: By existence of theorems of solutions to differential equations, y is unique for any given β and ϕ, together with ϕ_t, is in C^1 with respect to β for some region R in R^{n+1}.

Lemma 1: *All arcs in a field are stationary.*

Proof: Recall that a necessary condition for a line integral to be independent of path is that the cross partials be equal. Since the Hilbert integral is independent of path, by definition of a field we have:

(1) $\dfrac{\partial}{\partial y_i} f_{\dot{y}_k}(t,y,q) = \dfrac{\partial}{\partial y_k} f_{\dot{y}_i}(t,y,q)$, and

(2) $\dfrac{\partial}{\partial y_k} (f(t,y,q) - q \cdot f_{\dot{y}_k}) = \dfrac{\partial}{\partial t} f_{\dot{y}_k}(t,y,q)$.

Performing the differentiation in (2) we have:

(3) $f_{y_k} + (f_{\dot{y}}) \cdot q_{y_k} - q_{y_k} f_{\dot{y}} - \sum_i q_i \dfrac{\partial}{\partial y_k} f_{\dot{y}_i} = f_{\dot{y}_k t} + \sum_i f_{\dot{y} \dot{y}_i} \dfrac{\partial q_i}{\partial t}$ where

q_{y_k} is the n-vector of partial derivatives of q with respect to y_k.

Thus, by (1):

$$(4)\quad f_{y_k} = f_{\dot{y}_k t} + \sum_i q_i \frac{\partial}{\partial y_i} f_{\dot{y}_k} + \sum_i f_{\dot{y}\dot{y}_i} \frac{\partial q_i}{\partial t}$$

But $\frac{\partial}{\partial y_i} f_{\dot{y}_k} = f_{\dot{y}_k y_i} + \sum_j f_{\dot{y}_k \dot{y}_j} \frac{\partial q_j}{\partial y_i}$, and

$$(5)\quad f_{y_k} = f_{\dot{y}_k t} + \sum_i f_{\dot{y}_k \dot{y}_i} q_i + \sum_i f_{\dot{y}_k \dot{y}_i} \left(\frac{\partial q_i}{\partial t} + \sum_j j \frac{\partial q_i}{\partial y_j} \right)$$

But, (5) holds for $\dot{y}_i = q_i$ and for

$$(6)\quad \ddot{y}_i = \frac{\partial q_i}{\partial t} + \sum_j \frac{\partial q_i}{\partial y_j} \frac{dy_j}{dt} = \frac{\partial q_i}{\partial t} + \sum_j \frac{\partial q_i}{\partial y_j} q_j.$$

By (5) and (6) we have:

$$f_{y_k} = f_{\dot{y}_k t} + \sum_i f_{\dot{y}_k \dot{y}_i} \ddot{y}_i = \frac{d}{dt} f_{\dot{y}_k}.$$

□Lemma: *Let* S(B) *be an n-parameter family of stationary arcs for* J *such that:*

1) $y = \phi(t,\beta)$, $\phi: R \to D$,
2) $|\phi_\beta| \neq 0$,
3) R is simply connected.

Then $\langle S(B), \dot{y} \rangle$ *is a field if and only if:*

$$4)\quad \sum_{i=1}^{n} \frac{\partial y_i}{\partial \beta_s} \frac{\partial v_i}{\partial \beta_r} - \frac{\partial y_i}{\partial \beta_r} \frac{\partial v_i}{\partial \beta_s} \overset{t}{\equiv} 0,$$

for r,s *going from 1 to n and where* $v_i = f_{\dot{y}_k t}(t,y,\dot{y})$.

Proof: The implicit function theorem applies and we may solve 1) for β and get:

$$\beta_l = \psi_i(t,y)$$

We shall show that the slope of the field is $\dot{y} = \phi_t(t,\beta) = \phi_t(t,\psi(t,y)) = q$. We shall show that this choice of q is proper if and only if 4) holds by showing that 4) is necessary and sufficient for the Hilbert integral to be independent of the path of integration. Let L be an arc in S.

$$I_L = \int_L [(f(t,\phi,\phi_t) - q\cdot f_{\dot{y}}(t,\phi,\phi_t))\,dt + f_{\dot{y}}d\phi]$$

$$= \int_L (f + \phi_t\cdot f_{\dot{y}})dt - \int_L f_{\dot{y}}\phi_t\,dt + \int_L f_{\dot{y}}\,d\phi$$

$$= \int_L fdt + \int_L f_{\dot{y}}\cdot\phi_t dt - \int_L f_{\dot{y}}\cdot\phi_t dt + \int_L f_{\dot{y}}\cdot d\phi$$

$$= \int_L fdt + \sum_k \sum_i f_{\dot{y}_i}\,\omega^i_{\beta_k} d\beta_k.$$

Since R is simply connected, a necessary and sufficient condition for the last integral to be independent of path is that the cross partials of the integral be equal i.e., that:

(1) $\quad \dfrac{\partial}{\partial\beta_k} f = \dfrac{d}{dt} f_{\dot{y}_i}\,\phi^i_{\beta_k}$ and

(2) $\quad \dfrac{\partial}{\partial\beta_r}\Big(\sum_i f_{\dot{y}_i}\,\phi^i_{\beta_s}\Big) = \dfrac{\partial}{\partial\beta_s}\Big(\sum_i f_{\dot{y}_i}\,\phi^i_{\beta_r}\Big).$

Performing the differentiation in (1) we have:

(1)' $\quad \sum_i f_{y_i}\phi^i_{\beta_k} + \sum_i f_{\dot{y}_i}\,\phi^i_{t\beta_k} = \sum_i f_{\dot{y}_i}\,\omega^i_{\beta_{kt}} + \sum_i \omega^i_{\beta_k}\dfrac{d}{dt} f_{\dot{y}_i},$

which is equivalent to:

(1)'' $\quad \sum_i (f_{y_i} - \dfrac{d}{dt} f_{\dot{y}_i})\phi^i_{\beta_t} = 0,$

which is true by the stationarity of L. Thus independence of path is equivalent to (2), which may be written as:

$$(2)' \quad \sum_i \phi^i_{\beta_s} \frac{\partial}{\partial \beta_r} f_{\dot{y}_i} = \sum_i \phi^i_{\beta_r} \frac{\partial}{\partial \beta_s} f_{\dot{y}_i}.$$

Noting that $\phi^i_{\beta_j} = \frac{\partial y_i}{\partial \beta_i}$, we conclude that (2)' is the relation 4) in the lemma and proof is complete. □

Note: The terms in 4) are known as Lagrange brackets.

Theorem 5. (Sufficient condition of Weierstrass) *If* 1) $f \in C^2$. 2) $\bar{y}$ *is stationary.* 3) $\bar{y}$ *can be imbedded in a field, i.e. there exists a set* D *containing* $\bar{y}$ *that is covered by a field with* $\bar{y}$ *as one of its trajectories.* 4) $E(t,y, q(t,y), p) \leq 0$ *on* D, *with* p *finite and* $y \in D$. *Then* $\bar{y}$ *provides* J[y] *with a local maximum on* D.

Proof: Consider $\tilde{y} \in D \cap M$. To show: $J[\tilde{y}] - J[\bar{y}] \leq 0$. Along $\bar{y}$ we have, the Hilbert integral, $I_{\bar{y}} = \int_{\bar{y}} (f,(t,\bar{y},\dot{\bar{y}}) - \dot{\bar{y}} \cdot \bar{f}_{\dot{y}})\, dt - J[\bar{y}]$. Since $\tilde{y}$ is in the field, $I_{\tilde{y}}$ is independent of path, and

$$J[\bar{y}] = I_{\bar{y}} = I_{\tilde{y}} = \int_a^b [f(t,y,\dot{y}) - (q(t,y) - \dot{y}) \cdot f_{\dot{y}}(t,y,q(t,y))]dt.$$

Thus

$$J[y] - J[\bar{y}] = \int_a^b [f(t,y,\dot{y}) - f(t,y,q(t,y) - (\dot{y} - q(t,y)f_{\dot{y}}(t,y,q(t,y))]dt,$$

which is non-positive by assumption 4) of the theorem. □

Theorem 6. (Legender sufficient condition.) *If* f *and* $\bar{y}$ *are as in theorem 5 and if* $\eta \bar{f}_{\dot{y}\dot{y}} \eta^* < 0$ *for* $\eta \neq 0$, *then* $\bar{y}$ *provides a weak local maximum for* J[y].

Proof: The proof consists of showing that there exists a weak neighborhood of $\bar{y}$ such that the values of the arcs in that neighborhood are in D and such that the Weierstrass inequality holds there. The theorem, then, follows from theorem 5.

□

1.6 PROBLEMS WITH VARIABLE END POINTS

The above review assumed that the end points $y(a)$ and $y(b)$ are not choice variables. By way of illustrating the concept of first variation, we now allow the end points to vary, i.e. we let the initial time t_0 and position $y(t_0) = y_0$ and the terminal time t_1 and position $y(t_1) = y_1$, we allow all of these to be choice variables.

Further we consider a maximand of the form:

$$(1) \qquad I(y) = \int_{t_0}^{t_1} f(t, y, \dot{y})\, dt + g(t_0, y_0, t_1, y_1).$$

As is usual in deriving necessary conditions for extrema, we start from a point which solves our problem and consider a "near by" point to get conditions on the "increments" of our maximand. Start from a point which maximizes $I[y]$, say $\hat{z}$ where $\hat{z} = (\hat{t}_0, \hat{y}_0, \hat{y}, \hat{\dot{y}}, \hat{t}_1, \hat{y}_1)$, and compute the change in I when we move from $\hat{z}$ to $z = \hat{z} + \Delta z$. We let

$$\Delta z = (st_0, sy_0, \eta(t), \dot{\eta}(t), st_1, sy_1),$$

where $\eta(t)$ is sectionally smooth. Now ΔI is given by

$$\Delta I = \int_{t_0+st_0}^{t_1+st_1} f(t, \hat{y}+\eta, \hat{\dot{y}}+\dot{\eta})\, dt +$$

$$+ g((\hat{t}_0 + \delta t_0, \hat{y}_0, \hat{y} + \delta y_0, \hat{t}_1 + \delta t_1, \hat{y}_1 + \delta y_1)$$

$$- \int_{\hat{t}_0}^{\hat{t}_1} f(\hat{t}, \hat{y}, \dot{\hat{y}})\, dt - g(\hat{t}_0, \hat{y}_0, \hat{t}_1, \hat{y}_1).$$

The graph below may help in motivating further calculations.

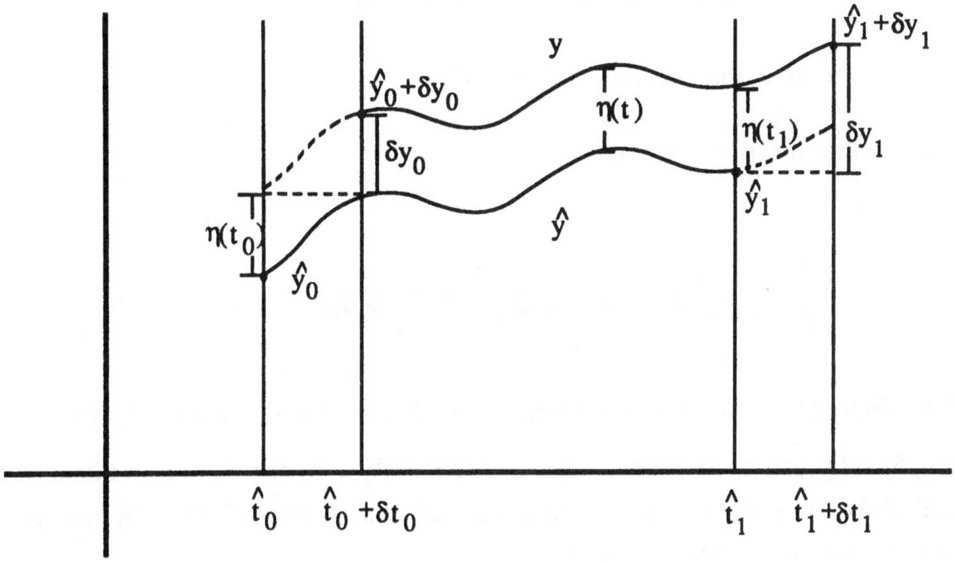

Figure 1. End point functionals.

First note that $\eta(t_0)$ and $\eta(t_1)$ can be approximated as follows:

(2.1) $\eta(t_0) \approx \delta y_0 - \dot{y}(t_0)\delta t_0$

(2.2) $\eta(t_1) \approx \delta y_1 - \dot{y}(t_1)\delta t_1$

Next we write:

$$\Delta I = \int_{\hat{t}_0}^{\hat{t}_1} [f(t, \hat{y} + \eta, \dot{\hat{y}} + \dot{\eta}) - f(\hat{t}, \hat{y}, \dot{\hat{y}})]\, dt +$$

$$+ \int_{\hat{t}_1}^{\hat{t}_1 + \delta t_1} f(t, \hat{y} + \eta, \dot{\hat{y}} + \dot{\eta})\, dt - \int_{\hat{t}_0}^{\hat{t}_0 + \delta t_0} f(t, \hat{y} + \eta, \dot{\hat{y}} + \dot{\eta})\, dt +$$

$$+ g(\hat{t}_0 + \delta t_0, \hat{y}_0 + \delta y_0, \hat{t}_1 + \delta t_1, \hat{y}_1 + \delta y_1) - g(\hat{t}_0, \hat{y}_0, \hat{t}_1, \hat{y}_1)$$

We approximate ΔI by:

$$\Delta I \approx \int_{\hat{t}_0}^{\hat{t}_1} [\hat{f}_y \eta + f_{\dot{y}} \dot{\eta}]\, dt + f(\hat{t}_1, \hat{y}_1, \hat{\dot{y}}_1)\, \delta t_1$$

$$- f(\hat{t}_0, \hat{y}_0, \hat{\dot{y}}_0)\, \delta t_0 + \hat{g}_{t_0} \delta t_0 + \hat{g}_{t_1} \delta t_1 + \hat{g}_{y_0} \delta y_0 + \hat{g}_{y_1} \delta y_1 =$$

$$= \delta I = \text{the variation of } I.$$

integrating by parts we get:

$$\delta I = \int_{\hat{t}_0}^{\hat{t}_1} (\hat{f}_y - \tfrac{d}{dt} f_{\dot{y}}) \eta\, dt + f_{\dot{y}} \eta \Big|_0^1 + f \delta y \Big|_0^1 + dg.$$

The first order necessary condition is that $\delta I \equiv 0$ identically in δt_0, $\delta t_1, \delta y_0, \delta y_1, \eta, \dot{\eta}$ as long as η is sectionally smooth, see e.g. Liusternik and Sobolev (1951). If we make use of 2 and take all the "increments" except one to be zero we get:

$$\text{(3.1')} \qquad \int_{\hat{t}_0}^{\hat{t}_1} [f_y - \tfrac{d}{dt} f_{\dot{y}}] \eta\, dt = 0 \quad \text{for all } \eta$$

$$\text{(3.2)} \qquad \hat{g}_{t_0} - f(\hat{t}_0, \hat{y}_0, \hat{\dot{y}}_0) + \hat{\dot{y}}_0 \hat{f}_{\dot{y}}(\hat{t}_0, \hat{y}_0, \hat{\dot{y}}_0) = 0$$

$$\text{(3.3)} \qquad \hat{g}_{t_1} + f(\hat{t}_1, \hat{y}_1, \hat{\dot{y}}_1) - \hat{\dot{y}}_1 \hat{f}_{\dot{y}}(\hat{t}_1, \hat{y}_1, \hat{\dot{y}}_1) = 0$$

$$\text{(3.4)} \qquad \hat{g}_{y_0} - \hat{f}_{\dot{y}}(\hat{t}_0, \hat{y}_0, \hat{\dot{y}}_0) = 0$$

$$\text{(3.5)} \qquad \hat{g}_{y_1} + \hat{f}_{\dot{y}}(\hat{t}_1, \hat{y}_1, \hat{\dot{y}}_1) = 0$$

By Dubois Reymond's lemma, see Gelfand and Fomin (1963), (3.1') Implies:

(3.1) $\frac{d}{dt} f_{\dot{y}} = f_y$

which is the Euler equation. Equations (3.2) - (3.5) are *the transversality conditions.* These conditions inherit their name from geometry. In fact if we think of the end points as restricted to lie on two curves, one at either end, then the transversality conditions state that a maximal arc is transversal to both end curves. For in instance suppose $y_0 = h(t_0)$ and $y_1 = r(t_1)$ represent such end curves and suppose the maximand is:

$$I[y] = \int_{t_0}^{t_1} f dt = \int_{t_0}^{t_1} \gamma(t,y) \sqrt{1 + \dot{y}^2}\, dt,$$

where γ is a real valued function. In this case $\delta y_0 = h'\delta t_0$ and $\delta y_1 = r'\delta t_0$. The conditions on the first variation of $I[y]$ then have the form:

(i) $2 \frac{d}{dt} \dot{y} \gamma / \sqrt{1 + y^2} = \gamma' \sqrt{1 + \dot{y}^2}$

(ii) $f + f_{\dot{y}} \cdot (r' - \dot{y})\Big|_{t_1} = 0$

(iii) $f + f_{\dot{y}} \cdot (h' - \dot{y})\Big|_{t_0} = 0$

In our example, (ii) and (iii) yield

$$(\gamma \cdot \sqrt{1 + \dot{y}^2} + (\dot{y} \gamma / \sqrt{1 + \dot{y}^2}) \cdot (r' - \dot{y})\Big|_{t_1} = 0$$

$$(\gamma \cdot \sqrt{1 + \dot{y}^2} + (\dot{y} \gamma / \sqrt{1 + \dot{y}^2}) \cdot (h' - \dot{y})\Big|_{t_0} = 0$$

And we then have:

$$(1 + r'\dot{y})\Big|_{t_1} = 0, \qquad (1 + r'\dot{y})\Big|_{t_0} = 0$$

So that

$$\dot{y}(t_1) = 1/r'(t_1), \qquad \dot{y}(t_0) = -1/h'(t_0)$$

The extremal y is perpendicular to both end curves, which certainly makes it transversal to them (since to be transversal all you have to do is to not come in or leave on a tangent i.e. penetrate and don't just touch). A more graphical illustration might help, see figure 2.

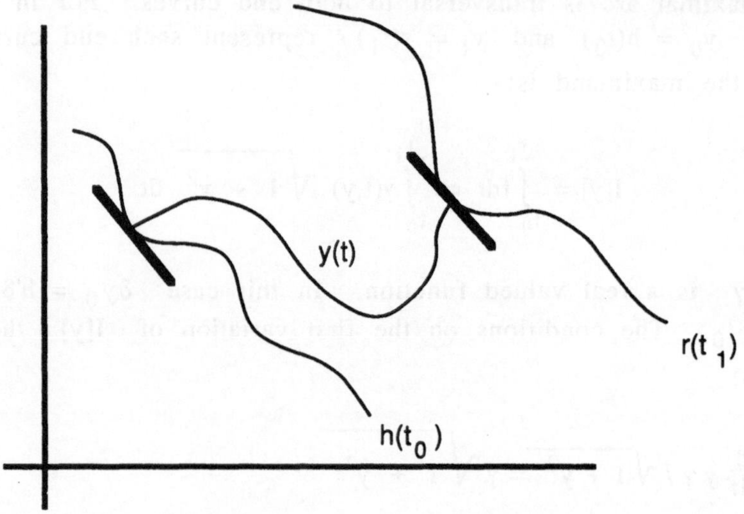

Figure 2. Transversality.

To help, at least as a memory aid, with the constrained problems that follow we restate our first order conditions in terms of canonical variables or Hamiltonian momenta. Define $P(t) = -f_{\dot{y}}$ and $H = f + p\dot{y}$. Then condition (3) can be written as

(3.1) $\dot{p} = -\hat{H}_y, \qquad \hat{\dot{y}} = \hat{H}_p$

(3.2) $\hat{g}_{t_0} - \hat{H}\Big|_{t_0} = 0$

(3.3) $\hat{g}_{t_1} + \hat{H}\Big|_{t_1} = 0$

(3.4) $\hat{g}_{y_0} + p(t_0) = 0$

(3.5) $\hat{g}_{y_1} - p(t_1) = 0$

1.7 EXAMPLES

a) **Cake eating:** So you have a cake x and some time T to eat it and you want to know how much to munch per second, so as to maximize your cumulative discounted utility. The problem is to maximize:

$$\int_0^T f(v)e^{-\rho t}\, dt \quad \text{with} \quad \dot{y} = -v,\ y(0) = x,$$

and $y(T) = 0$, where $y(t)$ is what remains of the cake by time t, and where v is the rate of consumption, and where ρ is the discount rate. The maximand can be written as

$$\int_0^T f(-\dot{y})\, e^{-\rho t}\, dt$$

Restricting oneself to interior solutions, Euler equation yields:

$$-f' e^{-\rho t} \equiv \text{constant} = c_1$$

or

$$f' = -c_1 e^{\rho t}.$$

Thus if $\rho = 0$, no discounting, then f' is constant over time and v is a constant and so $y(t) = c_2 t + c_3$. But $y(0) = x$ and $y(T) = 0$ so $c_3 = x$ and $c_2 = -x/t = -v$. If $\rho > 0$ then $\ddot{y} f'' = \rho f' e^{-\rho t} \geq 0$ since $f' \geq 0$. Thus if we assume diminishing marginal utility, $\dot{v} \geq 0$. Which confirms our prejudice; under discounting one eats the cake now rather than later but one spreads it evenly if the discount rate is zero.

b) **Dynamics of Monopoly:** A monopolist who does his own financing, can expand or contract at will. Say he uses capital k and labor ℓ to produce $x = f(k,\ell)$. Stock of capital changes according to:

$$\dot{k} = I - \alpha k,$$

where I is new machines and α is depreciation rate. Stock of labor changes according to

$$\dot{\ell} = y - \beta\ell,$$

where u is net new hiring and where β is natural attrition by retirement, death, and pestilence. The monopolist faces a demand curve where the quantity sold depends on the price of output, p, so that the inverse demand function is given by $p = h(x)$. Let the life time contract of workers by price at w and let machines cost q per machine. Then the discounted net cash flow will be:

$$(1) \qquad \pi = \int_0^T [h(x)\, x - wy - qI]\, e^{-rt}\, dt + q_T k_T\, e^{-\rho T}.$$

where r is a rate of discount and where q_T is the scrap value of the machines. Substituting into (1) we get:

$$(2) \qquad \pi = \int_0^T [hf - w\dot{\ell} - q\dot{k} - w\beta\ell - q\alpha k]\, e^{-rt}\, dt + q_T k_T\, e^{-rt} + w_T\, \ell_T\, e^{-rt}.$$

The gentle reader is not to think of $w_T \ell_T$ as the scrap value of workers but rather the value of the contracts which accrues to the firm when it shuts down and workers are let go.

The first order necessary conditions are

1) $\dot{w} + (\beta - r)w =$ value marginal product of labor

2) $\dot{q} + (\alpha - r)q =$ value marginal product of capital

3) $[h \cdot x - wu - qI]_T = s[q_T K_T + w_T L_T]$

Condition 3) provides a "natural boundary" condition for the death of the monopolistic firm. It dies when the profits from running the firm equals the interest on the value of it's assets.

c) **Consumption Over Time:** A consumer decides about a consumption vector $c(t)$ with c in E^n over a horizon $[0,T]$ which is fixed. Income at time t is $y(t)$ and the price vector is $p(t)$.

Consumers can borrow against future income but at time T he neither has debts or positive wealth so that $w(T) = 0$. As he earns money and spends it his wealth position changes according to:

$$\dot{w} = y - p \cdot c + rw,$$

where r is the interest rate (same for lending and borrowing). The instantaneous utility is $\varphi(c)$ and the utility functional is given by

$$U = \int_0^T \varphi(c)\ e^{-\rho t}$$

where ρ is the discount rate. Substituting for one of the goods, say c_n, from the wealth equation we have:

$$U = \int_0^T \varphi(c_1, \ldots, c_{n-1}, (y + rw - \dot{w} - \sum_{i=1}^{n-1} P_i c_i)/P_n)\ e^{-\rho t}\ dt.$$

The Euler equation yields:

i) $\varphi_i = (\varphi_n)\ P_i/P_n$

ii) $\frac{d}{dt}\left(\frac{\varphi_n}{P_n} e^{-\rho t}\right) = -r \frac{\varphi_n}{P_n}\ e^{-\rho t}$

In other words

iii) $\varphi_i = \gamma(P_i\ e^{(\rho - r)t}, \qquad i = 1, \ldots, n,$

where γ is a constant. In the special case where $\varphi = \sum_i \alpha_i \log c_i$, the Euler equations imply:

(*) $$\alpha_i = \gamma\ P_i\ x_i\ e^{(\rho - r)t}, \qquad i = 1, \ldots, n.$$

Summing over i and substituting in the wealth equation we get

$$\dot{w} = y - \frac{1}{\gamma} \sum \alpha_i\ e^{(r-\rho)t} + rw, \qquad w(T) = 0$$

Solving for w we get

$$w = e^{rt}[w_0 + \int_0^t y\ e^{-rs}\ ds - \frac{\Sigma\ \alpha_i}{\gamma}\int_0^t e^{-\rho s}\ ds]$$

$$= e^{rt}[w_0 + \int_0^t e^{-rs}\ ds + (e^{-\rho t} - 1)\frac{\Sigma\ \alpha_i}{\gamma}]$$

But since $w(T) = 0$ we have

$$\frac{1}{\gamma} = \frac{e^{rt}}{(\sum_i \alpha_i)(1 - e^{-\rho t})}(w_0 + \int_0^T y\ e^{-rt}\ dt)$$

Substituting in (*) and solving for x_i we get:

$$C_i = \left(\frac{\alpha_i}{\Sigma\ \alpha_i}\right)\left(\frac{e^{rt-\rho t}}{P_i e^{-rt}(1 - e^{-\rho t})}\right)(w_0 + \int_0^T y\ e^{-rt}\ dt)$$

Thus the current consumption expenditure, $\sum_i P_i\ C_i$ is given by

$$P \cdot C = \frac{e^{(r-\rho)t}}{e^{rt} - e^{(\rho-r)t}}(w_0 + \int_0^T y\ e^{-rt}\ dt)$$

In this last equation, the term in brackets is the "permanent income" of the consumer and we have shown that the permanent income hypothesis can be derived from consumers' utility maximization. is a Fisherian equilibrium so that $r = \rho$ then consumption is a fixed proportion of the permanent income which is the stronger form of the permanent income hypothesis.

2. STATEMENT OF THE PROBLEM OF BOLZA

Let T be a subset of the non-negative half line. Let M be the set of sectionally smooth functions from T into R^n.

<u>Problem A</u>: Among the arcs in M, maximize

(1) $J^0[y] = h^0(t_0), t_1, y_0, y_1) + \int_{t_0}^{t_1} L^0(t, y, \dot{y})\, dt$, where $t_0 < t_1$ are in T, $y_0 = y(t_0)$, $y_1 = y(t_1)$, subject to:

(2) $\varphi^\alpha(t, y, \dot{y}) = 0, \qquad \alpha = 1, \ldots, m, \; m < n,$

(3) $h^\beta(t_0, t_1, y_0, y_1) = 0, \qquad \beta = 1, \ldots, p, \; p < 2n + 1.$

We shall say that $\bar{y}$ provides a global, weak local or strong local solution to problem A if $\bar{y}$ provides a global, weak local or strong local maximum of (1) subject to (2) and (3).

The problem is equivalent* to the following problem:

<u>Problem A'</u>: Maximize $J^0[y]$, where $J^0[y]$ is defined in (1), subject to (2) and to

(3)' $J^\beta[y] = h^\beta(t_0, t_1, y_0, y_1) + \int_{t_0}^{t_1} L^0(t, y, \dot{y})\, dt.$

Problem A is a special case of problem A', as could be seen by setting $L^\beta = 0$ in (3)'. Problem A' is a special case of problem A, as could be seen by introducing the auxiliary variable $z^\beta(t)$ as follows:

(2)' $\dot{z}^\beta - L^\beta(t, y, \dot{y}) = 0,$

augmenting the constraints (2) by (2)' and writing (3)' as:

$$h^\beta(t_0, t_1, y_o, y_1) + z^\beta(t_1) - z^\beta(t_0) = 0$$

which is in the form of (3).

We shall outline the proofs of some of our theorems for problem A and state corresponding theorems for problem A', indicating the proof for some of these theorems.

We now introduce some definitions and notation. The variation of an arc y, due to a change in a parameter b, where b is a real

(*) In the sense that each problem may be formulated as a special case of the other.

vector with one or more components is denoted by δy and the variations of the points t_0 and t_1 are denoted, respectively by δt_0 and δt_1. The variation along an arc is a "shift" in the arc. Let $\bar{\Gamma} = (\bar{t}_0, \bar{t}_1, \bar{y})$ be a solution to problem A. We shall denote the variation along $\bar{\Gamma}$ by $p = (\tau_0, \tau_1, \eta)$, where p is an $n + 2$ vector if b has one component and p is an $s \times n + 2$ matrix if b has S components. In case p is a matrix, we shall denote the rows of p by $p^\sigma, \sigma = 1, ..., S$.

<u>Definition</u>: <u>An s-parameter family of arcs</u>. Let b be as above and let $t_0(b)$ and $t_1(b)$ be real valued functions of b. Let $y(t,b)$ be an n-vector valued function of b for $t \in [t_0(b), t_1(b)]$. We say that $y(t,b)$ is an s-parameter family of arcs.

<u>Definition</u>: <u>Admissible representations</u>. An s-parameter family of arcs is said to be an admissible representation starting at $\bar{y}$ if:

1) $y(t,0) = \bar{y}$, $t_0(0) = \bar{t}_0$, $t_1(0) = \bar{t}_1$.
2) i) $\delta t_0 = t_{0b}(0)\, db = \tau_0\, db$
   ii) $\delta t_1 = t_{1b}(0)\, db = \tau_1\, db$
   iii) $\delta y_i = y_{ib}(t, 0)\, db = \eta_i(t)\, db$,

where the second subscript denotes differentiation with respect to b and where db is an s-vector.

3) $\phi^\alpha(t, y(t,b), \dot{y}(t,b)) = 0$
4) $\phi^\alpha(t, \eta, \dot{\eta}) = \omega^\alpha_y \eta^* + \omega^\alpha_{\dot{y}} \dot{\eta}^* = 0$,

where η is an $s \times n$ matrix. Equations 4) are called the equations of variation along $\bar{y}$.

3. FIRST ORDER NECESSARY CONDITIONS

In this section we state the multiplier rule for the problem of Bolza and outline the proof. The outline consists of a sequence of assertions, whose proofs may be found, e.g., in Bliss (1930) or (1946).

Theorem 7. *If:* 1) h. L^0 *and* ϕ *are in* C^2. 2) $\bar{\Gamma}$ *is a solution to problem A.* 3) *The matrices* $[\bar{f}_{\dot{y}}]$ *and* $[h_{t_0}\ h_{t_1}\ h_{y_0}\ h_{y_1}]$ *have maximal rank. Then there exist constants* $(\lambda_0,\gamma) \neq 0$ *with* p + 1 *components,* $(\lambda_0, \lambda(t))$ *is never zero and:*

(1) Euler-Lagrange equations: $\frac{d}{dt}\bar{F}_{\dot{y}} = \bar{F}_y$ where $F = \lambda_0 L^0 + \lambda\phi$.

(2) $\frac{d}{dt}(\bar{F} - \dot{\bar{y}}\bar{F}_{\dot{y}}) = \frac{\partial}{\partial t}\bar{F}$.

(3) Transversality conditions:

i) $\bar{G}_{t_0} + \dot{\bar{y}}\,\bar{F}_{\dot{y}}\Big|_{t=t_0} - \bar{F}\Big|_{t=t_0} = 0$

ii) $\bar{G}_{t_1} - \dot{\bar{y}}\,\bar{F}_{\dot{y}}\Big|_{t=t_1} + \bar{F}\Big|_{t=t_1} = 0$

iii) $\bar{G}_{y_0} - \bar{F}_{\dot{y}}\Big|_{t=t_0} = 0$

iv) $\bar{G}_{y_1} + \bar{F}_{\dot{y}}\Big|_{t=t_1} = 0,$

where $G = \lambda_0 h^0 + \gamma h$, $\dot{\bar{y}}_r = \dot{\bar{y}}(t_r)$, (r = 0, 1), $\bar{F}\Big|_{t=t_1} = \bar{F}$ *with all argument evaluated at* t_r *and* $F_{\dot{y}}\Big|_{t=t_1}$ *is defined similarly.*

Outline of proof:

1. Condition (2) follows from condition (1).

2. Condition (1) is implied by:

(1)' There exists a constant n-vector such that $\bar{F}_{\dot{y}} = \int_{t_0}^{t_1} \bar{F}_y\, dt + C.$

3. Condition (3) is equivalent to:

$$\text{(3)}' \quad [(F - \dot{y}F_{\dot{y}})dt + dyF_{\dot{y}}]\Big|_0^1 + dG \equiv 0 \qquad \begin{matrix} dt_0 \\ dt_1 \\ dy_0 \\ dy_1 \end{matrix}$$

To see this consider the transformation:

$$dt_0 = \tau_0 db, \quad dt_1 = \tau_1 db, \quad dy_{i0} = \dot{y}_{i0}z_0 + \eta_{i0}db, \quad dy_{i1} = \dot{y}_{i1}z_1 + \eta_{i1}db.$$

4. By 1, 2, and 3 it suffices to prove (1)' and (3)'.

5. There exists a (p+1)-parameter family which is an admissible representation, starting at $\bar{y}$. This is accomplished by adjoining, to the constraints $\phi^\alpha = 0$, n - m equations $\phi^\alpha(t, y, \dot{y}) = z^\sigma$, where $z^\sigma(t)$ are auxiliary variables. This is done so that $\begin{bmatrix} \phi^\alpha_{\dot{y}} \\ \phi^\sigma_{\dot{y}} \end{bmatrix}$ has rank m. We then solve the system for $\dot{y} = \psi(t, y, z)$. Then consider: $\dot{y} = \Psi(t, y, z + b^\sigma\zeta)$, $\sigma = 1, ..., p + 1$. Solutions depend on t_0, y_0, y_1, and on b. $y = y(t, b)$ obtained this way is an admissible representation. Substituting into J[y] we have: b = 0 solves the following problem:

Maximize J(b) subject to $h^\beta(b) = 0$, $\beta = 1, ..., p$. This is a problem in R^n, and theorem 1 of chapter 2 applies. Thus, there exist constants $(\lambda_0, \gamma) \neq 0$ such that:

$$\text{(1)} \quad \frac{\partial}{\partial b^\sigma}(\lambda_0 J + \gamma h(b))|_{b=0} = 0.$$

6. Denoting $\frac{\partial}{\partial b}J$ evaluated at b = 0 by J_1, we have:

$$(2)\ \ J_1 = (\bar{h}^0_{t_0} + \bar{h}^0_{y_0}\,\dot{y}_0)\,\tau_0 + (\bar{h}^0_{t_1} + \bar{h}^0_{y_1}\,\dot{\bar{y}}_1)\,\tau_1 +$$

$$+ \bar{h}^{0'}_{y_0}\,\eta_0 + \bar{h}^0_{t_1}\,\eta_1 + L^0_1\,\tau_1 - L^0_0\,\tau_0 + \int_{\bar{t}_0}^{\bar{t}_1} (f_y v + f_{\dot{y}}\dot{\eta})\,dt.$$

7. Denoting $\frac{\partial}{\partial b} h^\beta$ evaluated at $b = 0$ by h^β_1, we have:

$$(3)\ \ h^\beta_1 = (\bar{h}^\beta_{t_0} + \bar{h}^\beta_{y_0}\,\dot{y})\tau_0 + (\bar{h}^\beta_{t_1} + \bar{h}^\beta_{y_1}\,\dot{y})\tau_1 + \bar{h}^\beta_{y_0}\eta_0 + \bar{h}^\beta_{y_1}\eta_1.$$

8. Recall that our variations satisfy:

$$(4.i)\ \ \Phi^\alpha = \phi^\alpha_y\,\eta + \omega^\alpha_{\dot{y}}\,\dot{\eta} = 0$$

$$(4.ii)\ \ \Phi^\sigma = \bar{\phi}^\sigma_y\,\eta + \bar{\phi}^\sigma_{\dot{y}}\,\dot{\eta} - \zeta^\sigma = 0,$$ where ζ^σ is the variation of z^σ.

9. Multiplying equations (4) by some functions $\ell^\alpha(t)$ and $\ell^\sigma(t)$ and integrating, we get:

$$(5)\ \ \int_{\bar{t}_0}^{\bar{t}_1} \ell^I\Phi^I + \ell^{II}(\Phi^{II} - \zeta^{II})\,dt = 0,$$ where the components of ℓ^I and Φ^I are ℓ^α and Φ^α and where the components of ℓ^{II} and Φ^{II} are ℓ^σ and Φ^σ.

10. Define $\tilde{F} = \lambda_0 L^0 + \ell\phi$, where $\ell = (\ell^I, \ell^{II})$ and $\phi = (\phi^I, \phi^{II})$. Take ℓ to be solutions of:

$$(6)\ \ \tilde{F}_{\dot{y}} = \int_{t_0}^{t} \tilde{F}_y\,dt + C,$$ where C is an arbitrary constant vector.

11. We may take $\lambda^\alpha(t) = \ell^\alpha(t)$.

12. $\ell^{II}(t) = 0$, and $\tilde{F} = F$. This would establish (1)'. The assertion is proved as follows: multiply (2) by λ_0 and add (5) to the right hand side, and use integration by parts, (1), (6) and the arbitrariness of ζ to show that $\ell^{II} = 0$.

13. The transversality condition (3)' follows from (1).

4. WEIERSTRASS AND SECOND ORDER NECESSARY CONDITIONS

Define $E(t, y, \dot{y}, \lambda_0, \lambda, p) = F(t, y, p, \lambda_0, \lambda) - F(t, y, \dot{y}, \lambda_0, \lambda) - (p - \dot{y}) F_{\dot{y}}(t, y, \dot{y}, \lambda_0, \lambda)$.

Theorem 8. (Weierstrass necessary condition). *If all the hypotheses of theorem 7 are satisfied then* $E(t,\bar{y}, \dot{\bar{y}}, p, \lambda_0, \lambda) \leq 0$ *for* $(\bar{y}, p)$ *satisfying* $\phi^{\alpha}(t, y, \dot{y}) = 0$.

Theorem 9. (Clebsch second order necessary conditions). *If all the hypotheses of theorem 7 are satisfied then* $\pi F_{\dot{y}\dot{y}} \pi^* \leq 0$ *for all* π *with* $\phi^{\alpha}_{\dot{y}}\pi^* = 0, \alpha = 1, \ldots, m$.

<u>Remark</u>: Theorems (8) and (9) were proved by McShane [36] using separation of convex sets.

Before we state our next theorem we define the concept of normality and state a necessary and sufficient condition for normality.

<u>Definition</u>: Γ, a solution to problem A, is said to be normal if it satisfies the first order necessary conditions with $\lambda_0 = 1$ and with a multiplier vector, (λ, γ).

Lemma. Γ *is normal if and only if it satisfies the rank condition in the sense of definition D.13 of Chapter 1.*

Theorem 10. (Jacobi-Mayer-Bliss necessary conditions) *If, in addition to the assumptions of theorem 7, the rank condition is satisfied, then there exist multipliers as in theorem 7 with* $\lambda_0 = 1$ *such that:*

$$\bar{J}_2 = d^2G(\bar{t}_0, \bar{t}_1, \bar{y}_1; \tau_0, \tau_1, \eta_0, \eta_1) + [(\bar{F}_t - \bar{F}_y\dot{y})\tau_r^2 + 2\bar{F}_y\eta(t_r)\tau_r]_{r=0}^{r=1}$$

$$+ \int_{\bar{t}_0}^{\bar{t}_1} \eta\bar{F}_{yy}\,\eta^* + 2\eta\,\bar{F}_{y\dot{y}}\dot{y}^* + \eta\bar{F}_{\dot{y}\dot{y}}\dot{\eta}^*\,dt \leq 0, \text{ for } (\tau_0, \tau_1, h) = p,\ \Phi^\alpha = 0,$$

and $$\bar{H}^\beta = (h^\beta_{t_0} + \dot{y}_0 h^\beta_{y_0})\tau_0 + h^\beta_{y_0}\eta_0 + (h^\beta_{t_1} + \dot{y}_1 h^\beta_{y_1})\tau_1 + h^\beta_{y_1}\eta_1 = 0.$$

5. SUFFICIENT CONDITIONS

Theorem 11. (First order sufficient conditions) *If* 1) L^0, h^0, h, *and* ϕ *are of class* C^2. 2) $\bar{\Gamma}$ *satisfies the first order necessary conditions, Euler-Lagrange Equations, with* $\lambda_0 > 0$. 3) *The functions* F *and* G, *defined in theorem 7 are concave. Then* $\bar{\Gamma}$ *is a global solution to problem A.*

Theorem 12. (Second order sufficient conditions) *If, in addition to assumptions* 1) *and* 2) *of theorem 11, we have* $\bar{J}_2 < 0$ *for every* p $\neq 0$ *with* $\Phi^\alpha = 0$ *and* $H^\beta = 0$. *Then* $\bar{\Gamma}$ *is a weak local solution of problem A.*

6. CHARACTERIZATION OF PROBLEM A'

(a) *Necessary conditions:* If 1) h^0, h, L^0, L and ϕ are of class C^2. 2) Γ is a solution to problem A'. 3) The matrices $\begin{bmatrix} \bar{\Phi}_{\dot{y}} \\ \bar{L}_{\dot{y}} \end{bmatrix}$, $[\bar{h}_{t_0}, \bar{h}_{t_1}, \bar{h}_{y_0}, \bar{h}_{y_1}]$ have maximal ranks. Then there exists constants $(\lambda_0, \gamma') \neq 0$ and functions $\lambda'(t)$ with (λ_0, γ') never zero such that:

(1) Euler-Lagrange Equations: $\frac{d}{dt} F_{\dot{y}} = F_{y'}$, where $F' = \lambda^0 L^0 + \gamma L + \lambda\phi$.

(2) $\frac{d}{dt}(F' - \dot{y}F'_{\dot{y}}) = \frac{\partial}{\partial t} F'$

(3) Transversality conditions: $[(F' - \dot{y}F'_y)\, dt + dyF'_{\dot{y}}]\Big|_0^1 + dG' \overset{dt_0,\, dt_0,\, dy_0,\, dy_1}{\equiv} 0,$

where $G' = \lambda_0 h^0 + \gamma' h$.

(4) Weierstrass condition. Define $E'(t, y, \dot{y}, p, \lambda_0, \lambda') = F(t, y, p, \lambda_0, \lambda') - F'(t, y, \dot{y}, \lambda_0, \lambda') - (p - \dot{y})F'_{\dot{y}}(t, y, \dot{y}, \lambda)$. $E(t, \bar{y}, \dot{\bar{y}}, p, \lambda_0, \lambda') \leq 0$ for y, p satisfying $\phi^\alpha(t, y, p) = 0$.

(5) Clebsch condition. $\pi \bar{F}'_{\dot{y}\dot{y}} \pi^* \leq 0$ for π satisfying $\bar{\Phi}_{\dot{y}} \pi^* = 0$.

(6) Mayer-Jacobi-Bliss condition. If Γ is normal then $J'_2 \leq 0$ for each p with $\Phi^\alpha = 0$, $H^\beta + \int_{t_0}^{\bar{t}_1} (L^\beta_y \eta + L^\beta_{\dot{y}} \dot{\eta})\, dt = 0$, where J'_2 is analogous to J_2 with F and G replaced by F' and G'.

(b) *Sufficient conditions:* Suppose h^0, h, L^0, L, and ϕ are of class C^2 and conditions 1) and 3) of (a) above hold at Γ. Then we have:

(b.1) If $\lambda_0 > 0$ and F' and G' are concave then Γ is a global solution to problem A'.

(b.2) If $J_2' < 0$ for every $p \neq 0$ with $\Phi^\alpha = 0$ and $H'^\beta = 0$, then Γ is a weak local solution to problem A'.

We introduce the auxiliary variables z^β as follows:

$$\dot{z}^\beta - L^\beta(t, y, \dot{y}) = 0, \qquad \beta = 1, \ldots, p.$$

Thus, we have a problem of type A of Maximizing J subject to $\phi^\alpha = 0$, $\dot{z}^\beta - L^\beta = 0$, $h^\beta + z^\beta_1 - z^\beta_0 = 0$. Define the, path, Lagrangian $F = \lambda_0 L^0 + \lambda^I \phi + \lambda^{II} L$ and the end point Lagrangian $G = \lambda_0 L^0 + \gamma h$. by the Euler-Lagrange conditions for problem A, $\frac{d}{dt} F_{\dot{y}} = F_y$ and $\frac{d}{dt} F_{\dot{y}} = F_y$ and $\frac{d}{dt} F_{\dot{z}} = F_z = 0$. But $F_{\dot{z}} = \lambda^{II}$, hence $\dot{\lambda}^{II} = 0$, and λ^{II} is a constant vector. By transversality conditions for problem a, $\lambda^{II} = \gamma$. Thus taking $\lambda' = \lambda^I$ and $\gamma' = \gamma$ the characterization of problem A' follows from that of problem A.

CHAPTER 6

THE PROBLEM OF BOLZA WITH EQUALITY-INEQUALITY CONSTRAINTS

In this chapter, we study the problem of Bolza with added inequality constraints. We shall use the theorems of chapter 5 to prove our characterization of the present problem: Problem A". Maximize J[y] subject to:

1. $\phi^{\bar{\alpha}}(t, y, \dot{y}) = 0, \quad \bar{\alpha} = 1, ..., m_1.$

2. $\phi^{\bar{\bar{\alpha}}}(t, y, \dot{y}) \geq 0, \quad \bar{\bar{\alpha}} = 1, ..., m.$

3. $h^{\bar{\beta}}(t_0, t_1, y_0, y_1), \ \bar{\beta} = 1, ..., p_1.$

4. $h^{\bar{\bar{\beta}}}(t_0, t_1, y_0, y_1), \ \bar{\bar{\beta}} = 1, ..., p.$

Let $\bar{\phi}$ and $\bar{\bar{\phi}}$ be vectors whose components are $\phi^{\bar{\alpha}}$ and $\phi^{\bar{\bar{\alpha}}}$ respectively and similarly define $\bar{h}$ and $\bar{\bar{h}}$ to be vectors whose components are $h^{\bar{\beta}}$ and $h^{\bar{\bar{\beta}}}$. Let $\phi = (\bar{\phi}, \bar{\bar{\phi}})$ and $h = (\bar{h}, \bar{\bar{h}})$. Define the set $A = \{1, ..., m_1\} \cup \{m_1 + 1 \leq \bar{\bar{\alpha}} \leq m : \omega^{\alpha}(t, \hat{y}, \dot{\hat{y}}) = 0$, for a given $\hat{\Gamma}\}$, and the set $B = \{1, ..., p_1\} \cup \{m_1 + 1 \leq \bar{\bar{\beta}} \leq m : h^{\bar{\bar{\beta}}}(\hat{t}_0, \hat{t}_1, \hat{y}_0, \hat{y}_1) = 0$, for a given $\hat{\Gamma}\}$, where $\hat{\Gamma} = \hat{t}_0, \hat{t}_1, \hat{y}_0, \hat{y}_1$. The sets A and B are indices of constraints that are equality constraints or that are effective at $\hat{\Gamma}$. Let a = the number of elements in A and let b be the number of elements in B.

In section 1 we discuss necessary conditions and in section 2 we discuss sufficient conditions.

1. NECESSARY CONDITIONS

Theorem 1. *If* 1) L^0, ϕ, *and* h *are of class* C^2. 2) $\hat{\Gamma}$ *is a solution to problem A''.* 3) *The matrices* $[\hat{\phi}^{\alpha}_{\dot{y}}]$, $\alpha \in A$ *and* $[h^{\beta}_{t_0}, h^{\beta}_{t_1}, h^{\beta}_{y_0}, h^{\beta}_{y_1}]$, $\beta \in B$ *have ranks* a *and* b *respectively. Then there exist constants* $(\lambda_0, \bar{\gamma} \cdot \bar{\bar{\gamma}}) \neq 0$, $\lambda_0 \geq 0$ *and functions* $(\bar{\lambda}, \bar{\bar{\lambda}})$ *such that* $(\lambda_0, \bar{\lambda}, \bar{\bar{\lambda}}) \neq 0$ *at any* $t \in [\bar{t}_0, \bar{t}_1]$ *such that:*

(1.1) (a) $\bar{\bar{\lambda}} \geq 0$ *for each of its components,* $\lambda^{\bar{\bar{\alpha}}}$, *we have*

(b) $\lambda^{\bar{\bar{\alpha}}}\hat{\phi}^{\bar{\bar{\alpha}}} = 0.$

(1.2) (a) $\bar{\bar{\gamma}} \geq 0$ *for each of its components,* $\gamma^{\bar{\bar{\beta}}}$, *we have*

(b) $\gamma^{\bar{\bar{\beta}}}\hat{h}^{\bar{\bar{\beta}}} = 0.$

(1.3) *Euler-Lagrange-Valentine equations:*

$\frac{d}{dt}\hat{F}^2_{\dot{y}} = \hat{F}^2_{y}$, *where* $F^2 = \lambda_0 L^0 + \bar{\lambda}\bar{\phi} + \bar{\bar{\lambda}}\bar{\bar{\phi}}$.

(1.4) $\frac{d}{dt}(\hat{F}^2 - \dot{y}\hat{F}^2_{\dot{y}}) = \frac{\partial}{\partial t}\hat{F}^2$

(1.5) *Transversality conditions:*

i) $\hat{G}^2_{t_0} + \dot{y}\,\bar{F}^2_{\dot{y}}\Big|_{t=t_0} - \hat{F}^2\big|_{t=\hat{t}_0} = 0$

ii) $\hat{G}^2_{t_1} - \dot{y}\,\bar{F}^2_{\dot{y}}\Big|_{t=t_1} + \hat{F}^2\big|_{t=\hat{t}_1} = 0$

iii) $\hat{G}^2_{y_0} - \hat{F}^2_{\dot{y}}\big|_{t=\hat{t}_0} = 0$

iv) $\hat{G}^2_{y_1} + \hat{F}^2_{\dot{y}}\big|_{t=\hat{t}_1} = 0$, *where* $G^2 = \lambda_0 h^0 + \bar{\gamma}\bar{h} + \bar{\bar{\gamma}}\bar{\bar{h}}$.

(2) *Weierstrass.* $E^2(t, \hat{y}, \dot{\hat{y}}, p, \bar{\lambda}, \bar{\bar{\lambda}}) \leq \bar{\bar{\lambda}}\bar{\bar{\phi}}(t, y, p)$, *for every* $(\hat{y}, p)$ *satisfying the constraints* 1. *and* 2., *where* $E^2 = F^2(t, y, \dot{y}, \lambda_0, \lambda) - (p - \dot{y}) F^2_{\dot{y}}(t, y, \dot{y}, \lambda_0, \lambda)$.

(3) *Clebsch.* $\pi\hat{F}^2_{\dot{y}\dot{y}}\pi^* \leq 0$ for all π with $\hat{\phi}^{\alpha}_{\dot{y}}\pi^* = 0, \alpha \in A$.

(4) *Jacobi-Mayer-Bliss: If* $\hat{\Gamma}$ *is normal then* λ_0, *in* 1., 2., *and* 3., *is* 1 *and:*

$$J_2^2 = d^2G(\tau_0, \tau_1, \eta_0, \eta_1) + [(\hat{F}_t^2 - \dot{y}\hat{F}_y^2)(\tau_r)^2 + 2\hat{F}_y^2\eta(r_r)\tau_r]_{r=0}^{r=1} +$$

$$+ \int_{t_0}^{t_1} [\eta\hat{F}_{yy}^2\eta^* + 2\eta\hat{F}_{y\dot{y}}^2\eta^* + \dot{\eta}\hat{F}_{\dot{y}\dot{y}}^2\dot{\eta}^*]\, dt \leq 0$$

for p *satisfying* $\Phi^\alpha = 0$ *and* $\hat{H}^\beta = 0, \alpha \in A, \beta \in B.$

Outline of proof: We shall outline the proof of the above theorem. Let us introduce two auxiliary vectors of variables $z(t)$ and $\omega(t)$ with $m - m_1$ and $p - p_1$ component respectively. We may rewrite the constraints 2. as:

(2') $\phi^{\bar{\alpha}} - (\dot{z}^{\bar{\alpha}})^2 = 0$, $z^{\bar{\alpha}}(t_0) = 0$, $z^{\bar{\alpha}}(t_1)$ free and the constraints (4) as:

(4') $h^{\bar{\beta}} - \omega^{\bar{\beta}}(t_1))^2 = 0$, $\omega^{\bar{\beta}}(t_0)$ free.

Now our problem is to maximize J subject to (1), (2'), (3), and (4') which is a problem of type A. Define $F = \lambda_0 L^0 + \sum_{\bar{\alpha}} \lambda^{\bar{\alpha}}\phi^{\bar{\alpha}} +$

$\sum_{\bar{\alpha}} \lambda^{\bar{\alpha}}(\phi^{\bar{\alpha}} - (z^{\bar{\alpha}})^2)$ and $G = \lambda_0 L^0 + \sum_{\bar{\beta}} \gamma^{\bar{\beta}} h^{\bar{\beta}} + \sum_{\bar{\beta}} \gamma^{\bar{\beta}} h^{\bar{\beta}} (h^{\bar{\beta}} - (\omega^{\bar{\beta}}(t_1))^2).$

By the Euler-Lagrange equations for problem A, condition (1), theorem 7, chapter 5, we have:

(1.1) $\frac{d}{dt}\hat{F}_{\dot{y}} = \hat{F}_y.$

(1.2) $\frac{d}{dt}\hat{F}_{\dot{z}} = \hat{F}_z.$

But, $F_{\dot{z}^{\bar{\alpha}}} = -2\lambda^{\bar{\alpha}}\, \hat{\dot{z}}^{\bar{\alpha}}$ and $\hat{F}_{z^{\bar{\alpha}}} = 0$. Thus by (1.2),

(2) $\mu^{\bar{\alpha}}\, \hat{\dot{z}}^{\bar{\alpha}} = \text{constant}.$

By transversality conditions for problem A, conditions (3) of theorem 7, chapter 5, we have:

(3.1) $\hat{G}_{t_0} + (\dot{\hat{y}}\hat{F}_{\dot{y}} + \dot{\hat{z}}\hat{F}_{\dot{z}} + \dot{\hat{\omega}}\hat{F}_{\dot{\omega}} - \hat{F})|_{t=t_0} = 0$

(3.2) $\hat{G}_{t_1} + (\hat{F} - \dot{\hat{y}}\hat{F}_{\dot{y}} + \dot{\hat{z}}\hat{F}_{\dot{z}} + \dot{\hat{\omega}}\hat{F}_{\dot{\omega}})|_{t=t_1} = 0$

(3.3.1) $\hat{G}_{y_0} - \hat{F}_{\dot{y}}|_{t=\hat{t}_0} = 0.$

(3.3.2) $\hat{G}_{\omega_0} - \hat{F}_{\dot{\omega}}|_{t=\hat{t}_0} = 0.$

(3.4.1) $\hat{G}_{y_1} - \hat{F}_{\dot{y}}|_{t=\hat{t}_1} = 0.$

(3.4.2) $\hat{G}_{z_1} - \hat{F}_{\dot{z}}|_{t=\hat{t}_1} = 0.$

(3.4.3) $\hat{G}_{\omega_1} - \hat{F}_{\dot{\omega}}|_{t=\hat{t}_1} = 0.$

By definition of F and G we have $\hat{G}_{\omega_0} = 0$, $\hat{F}_{\dot{\omega}} = 0$, $\hat{G}_{z_1} = 0$, $\hat{F}_{\dot{z}^{\bar{\alpha}}} = -2\mu^{\bar{\alpha}}\, \dot{z}^{\bar{\alpha}}$, $\hat{G}_{\omega_1^{\bar{\beta}}} = -2\gamma^{\bar{\beta}}\omega^{\bar{\beta}}$ and $\hat{F}_{\dot{\omega}} = 0$. Thus we write (3) as:

(4.1) $\hat{G}_{t_0} + \dot{\hat{y}}\hat{F}_{\dot{y}}|_{t=t_0} - \hat{F}|_{t=t_0} = 0$

(4.2) $\hat{G}_{t_1} + (\hat{F} - \dot{\hat{y}}\hat{F}_{\dot{y}})|_{t=t_1} = 0$

(4.3) $\hat{G}_{y_0} - \hat{F}_{\dot{y}}|_{t=\hat{t}_0} = 0.$

(4.4.1) $\hat{G}_{y_1} - \hat{F}_{\dot{y}}|_{t=\hat{t}_1} = 0.$

(4.4.2) $\hat{F}_{\dot{z}}|_{t=\hat{t}_1} = -2\lambda^{\bar{\alpha}}(t_1)\, \dot{z}^{\bar{\alpha}}(t_1) = 0$

i. Proof of 1: By (4.4.2) and (2) we have: $\mu^{\bar{\alpha}}\, \hat{\dot{z}}^{\bar{\alpha}} = 0$, which establishes (1.1)-b of the theorem, since $\hat{\dot{z}}^{\bar{\alpha}} \neq 0$ if and only if: $\hat{\phi}^{\bar{\alpha}} > 0$.

By (4.4.3) we have: $\gamma^{\bar{\beta}}\, \hat{h}^{\bar{\beta}} = 0$ which is condition (1.2)-b of the theorem.

Now take $\lambda^{\bar{\alpha}}$ to be the components of $\bar{\lambda}$ of the theorem and $\lambda^{\bar{\bar{\alpha}}}$ as the components of $\bar{\bar{\lambda}}$. Also take $\gamma^{\bar{\beta}}$ and $\gamma^{\bar{\bar{\beta}}}$ as the components of $\bar{\gamma}$ and $\bar{\bar{\gamma}}$. It is then obvious that the "derivatives", with respect to y, $\dot{y}$, t_0, t_1, y_0, and y_1 of F and G are equal to the derivatives of F^2 and G^2 along $\hat{\Gamma}$. This establishes (1.3), (1.4), and (1.5) of the theorem.

We now outline the proof of parts (1.1)-a and (1.2)-a of the theorem. Fix $\tilde{t} \in [\hat{t}_0, \hat{t}]$. If $\bar{\bar{\alpha}} \notin A$, then $\lambda^{\bar{\bar{\alpha}}} = 0$, by (1.1)-b. We now show that $\lambda^{\bar{\bar{\alpha}}} \geq 0$ for $\bar{\bar{\alpha}} \in A$. At $\tilde{t}$ we may, without loss of generality, ignore, locally, the

constraints $\phi^{\bar{\bar{\alpha}}}$ for $\bar{\bar{\alpha}} \notin A$. By the Clebsch condition for this problem of type A we have:

$$\pi_1 \hat{F}_{\dot{y}\dot{y}} \pi_1^* + \pi_1 \hat{F}_{\dot{y}\dot{z}} \pi_2^* + \pi_2 \hat{F}_{\dot{z}\dot{y}} \pi_1^* + \pi_2 \hat{F}_{\dot{z}\dot{z}} \pi_2^* \leq 0$$

for π_1, π_2 satisfying:

(5.1) $\quad \hat{\phi}^{\bar{\alpha}}\, \pi_1^* + \hat{\phi}^{\bar{\alpha}}_{\dot{z}}\, \pi_2^* = \hat{\phi}^{\bar{\alpha}}_{\dot{y}} = 0$

(5.2) $\quad \hat{\phi}^{\alpha}_{\dot{y}}\, \pi_1^* + \hat{\phi}^{\bar{\bar{\alpha}}}_{\dot{z}}\, \pi_2^* = 0, \qquad \alpha \in A$

But $\hat{\phi}^{\bar{\bar{\alpha}}}_{\dot{z}} = -2\hat{z}^{\bar{\bar{\alpha}}} = 0$, $\alpha \in A$. Thus (5) does not restrict π_2. Furthermore, $\pi_1 = 0$ satisfies (5). Also $\hat{F}_{\dot{y}\dot{z}} = 0$ and $\hat{F}_{\dot{z}\dot{y}} = 0$. Now take $\pi_1 = 0$ and $\pi_2 = 0$ except for one component $\bar{\bar{\alpha}}$. The inequality still holds and $\hat{F}_{\dot{z}\cdot\bar{\bar{\alpha}}\dot{z}\bar{\bar{\alpha}}}(\pi^{\bar{\bar{\alpha}}}) = -2\lambda^{\bar{\bar{\alpha}}}(\pi^{\bar{\bar{\alpha}}})^2 \leq 0$. Hence $\lambda^{\bar{\bar{\alpha}}} \geq 0$ for $\bar{\bar{\alpha}} \in A$.

We now indicate the proof of (1.2)-b. Going back to the outline of the proof of theorem 7, after reducing problem A to a finite dimensional problem. Noting that the first order necessary condition of chapter 3 applies, we conclude that $\gamma^{\bar{\bar{\beta}}} \geq 0$.

ii. Proof of 2: Recalling the definition of $F^{(1)}$ in (i), we have:

(6) $E(t, \hat{y}, \hat{z}, \hat{y}, \hat{z}, p, p, \lambda_0, \lambda)^{(2)}$

$$= F^2(t,\hat{y}, p) - \sum_{\alpha} \lambda^{\bar{\alpha}}(p^{\bar{\alpha}})^2 - \hat{F}^2 + \sum_{\bar{\bar{\alpha}}} \lambda^{\bar{\alpha}}(\hat{t}^{\bar{\alpha}})^2 - (p - y)F^2_{\dot{y}} - (p -$$

$z)F_{\dot{z}}$.

(1) $F = F^2 - \bar{\bar{\alpha}}\lambda^{\alpha}(z^{\bar{\alpha}})^2$

(2) p is the "comparison" vector of z.

The fourth and last terms of the right hand side of (6) are zeros, by i, and we have

$$E(t, \hat{y}, \hat{z}, \hat{y}, \hat{z}, p, p, \lambda_0, \lambda) = E^2(t, \hat{y}, \hat{y}, p, \lambda_0, \lambda) - \sum_{\alpha} \lambda^{\bar{\alpha}}(p^{\bar{\alpha}})^2.$$

By the Weierstrass condition for problem A, $E \leq 0$ for $(\hat{y}, p, p)$ satisfying constraints (1) and (2)', i.e., satisfying (1) and $(p^{\bar{\alpha}})^2 = \phi^{\bar{\alpha}}$. Thus $E^2(t, \hat{y}, \hat{y}, p, \lambda_0, \lambda) - \lambda\phi \leq 0$ as we were to show.

iii. Proof of 3: Condition (3) follows immediately from the Clebsch conditions for problem A, arguing as we did towards the end of i above.

iv. Proof of 4: By the second order necessary condition for problem A we have $\hat{J}_2 \leq 0$ for $\hat{\Phi} = 0$ and $\hat{H} = 0$. But $\hat{J}_2 = d^2\hat{G}^2 - 2\sum_{\bar{\bar{\beta}}} \gamma^{\bar{\beta}}(\omega^{\bar{\beta}})^2 + [(\hat{F}^2_t - \dot{y}\hat{F}^2_y)(\tau_r^2) + 2\hat{F}^2_y\eta(t_r)\tau_0]\Big|_{r=0}^{r=1} + \int_{\hat{t}_0}^{\hat{t}_1} \eta\hat{F}^2_{yy}\eta^* + 2\eta\hat{F}^2_{y\dot{y}}\dot{\eta}^* + \dot{\eta}\hat{F}^2_{\dot{y}\dot{y}}\dot{\eta}\, dt - \int_{\hat{t}_0}^{\hat{t}_1} \mu^{\bar{\alpha}}(\zeta^{\bar{\alpha}})^2$, where ω and ζ are the increments of w and z, since $F_t + F^2_t$, $F_z = 0$, $F_{\dot{z}z} = 0$. Thus:

$$\hat{J}_2 + \hat{J}_2 - \int_{\hat{t}_0}^{\hat{t}_1} \mu^{\bar{\alpha}}(\zeta^{\bar{\alpha}})^2,$$

since $\gamma^{\overline{\beta}}$ and $\mu^{\overline{\alpha}}$ are zeros if the corresponding constraints are ineffective. Thus the choice of "increments" $\omega^{\overline{\beta}}, \overline{\beta} \notin B$ and $\zeta^{\overline{\alpha}}, \notin A$ does not affect $\hat{J}_2$. Writing down $\Phi^{\alpha} = 0$, $\beta \in B$ we find that the choice of the remaining $\omega^{\overline{\beta}}$'s and $\zeta^{\overline{\alpha}}$'s is arbitrary since they have zero coefficients in the expressions. Thus we may chase them to be zeros and $\hat{J}_2 = \hat{J}_2^2$.

Finally we note that rank conditions and normality of problem A characterization are easily verified in view of the corresponding conditions for this problem.

2. SUFFICIENT CONDITIONS

Theorem 2. *Let* L^0, ϕ, *and* h *be of class* C^2 *and let* $\hat{\Gamma}$ *satisfy conclusion (1) of theorem 1 of this chapter with* $\lambda_0 > 0$. *Then:*

a) *If* ϕ^{α} *and* h^{β} *are linear and* $L^0, \phi^{\overline{\alpha}}, h^{\overline{\beta}}$ *are concave in* y *and* $\dot{y}$, *then* $\hat{\Gamma}$ *is a global solution of problem A".*
b) *If* $\hat{J}_2^2 < 0$ *for* $p \neq 0$ *with* $\hat{\phi}^{\alpha} = 0, \alpha \in A$ *and* $\hat{H}^{\beta} = 0, \beta \in B$, *then* $\hat{\Gamma}$ *is a weak solution to problem A".*

Part a) follows from the first order sufficient condition for problem A, since our hypothesis implies that the "path" and "endpoint" Lagrangians are concave. Part b follows from Pennissi's theorem, see Pennissi (1953), since our set of variations p with $\Phi^{\alpha} = 0$, $H^{\beta} = 0$, $\alpha \in A$, $\beta \in B$ includes his.

CHAPTER 7

EXTENSIONS AND APPLICATIONS

In this chapter we note some extensions of the theorems of chapter 6 and some applications. We first state a problem in optimal control and characterize its solutions as applications of chapter 6. The optimal control problem with scalar criterion is presented in section 1. In section 2 we present extensions of the control problem to: a) problems with time lags, b) problems with bounded state variables and, c) problems with finite vector criteria. For these problems we discuss only the first order necessary conditions and the Weierstrass conditions. We then present some economic applications.

1. AN OPTIMAL CONTROL PROBLEM

First we state an optimal control problem with scalar criterion and characterize it. Next we discuss normality. Then we discuss two special problems: a) when the controls are restricted to be non-negative and b) when the criterion is a function of integrals.

Suppose a dynamical system is given by:

(1) $\dot{x} = f(t, x, u)$,

where x is an n-vector function and u is an R vector function. x will be referred to as the state variable vector and u will be referred to as the control vector. Our problem is to maximize:

$$J[u] = g^0(t_0, t_1, x_0, x_1) + \int_{t_0}^{t_1} f^0(t, x, u)\, dt,$$

where $x_0 = x(t_0)$, $x_1 = x(t_1)$, subject to (1) and subject to:

(2.1) $\quad K^{\bar{\alpha}}(t, x, u) = 0, \qquad \bar{\alpha} = 1, ..., m_1.$

(2.2) $\quad K^{\bar{\bar{\alpha}}}(t, x, u) \geq 0, \qquad \bar{\bar{\alpha}} = m_1 + 1, ..., m.$

(2.3) $\quad g^{\bar{\beta}}(t_0, t_1, x_0, x_1) = 0, \qquad \bar{\beta} = 1, ..., p_1.$

(2.4) $\quad g^{\bar{\bar{\beta}}}(t_0, t_1, x_0, x_1) \geq 0, \qquad \bar{\bar{\beta}} = 1, ..., p_2.$

Note: We have suppressed the arguments of $x(t)$ and $u(t)$.

Let A be the set of indices α corresponding to either equality constraints (2.1) or to effective inequality constraints (2.2) at some arc $\hat{\Delta} = (\hat{\Delta}_1, \hat{\Delta}_2) = (\hat{t}_0, \hat{t}_1, \hat{x}_0, \hat{x}_1; \hat{x}, \hat{u})$. And let B be the set of indices β corresponding to equality constraints (3.1) or to effective inequality constraints (3.2).

As before, we say that an arc is normal if it is the first order necessary conditions with a unique vector of multipliers with $\lambda_0 = 1$. A necessary and sufficient condition for normality of a solution to the control problem will be stated below. Finally, denote the variation of $x(t)$ by $\xi(t)$, the variation of $u(t)$ by ν and let τ_0 and τ_1 be as before. Denote the variation vector by $S = (\tau_0, \tau_1, \xi_0, \xi_1, \xi, \nu)$.

Theorem 1. *Necessary conditions (Pontriagin-Hestenes-Berkovitz). If:* 1) $\hat{\Delta} = (\hat{t}_0, \hat{t}_1, \hat{x}_0, \hat{x}_1; \hat{x}, \hat{u})$ *be a solution to the control problem.* 2) *The functions* g^0, g, f^0, f, *and* K *are in class* C^2. 3) *The matrices* $[\hat{K}^{\alpha}_u]$, $\alpha \in A$ *and* $[\hat{g}^{\beta}_{\Delta_1}]$, $\beta \in B$ *have maximal ranks. Then there exist constants* $(\lambda_0, \bar{\gamma}, \bar{\bar{\gamma}})$, $\lambda_0 \geq 0$, *and time functions* $(\lambda(t), \bar{\mu}(t), \bar{\bar{\mu}}(t))$ *with* $(\lambda_0, \bar{\gamma}, \bar{\bar{\gamma}}) \neq 0$ and $(\lambda_0, \lambda, \mu) \neq 0$ *such that:*

1.i. $\quad \bar{\bar{\mu}} \geq 0$ *and* $\bar{\bar{\mu}}\hat{\bar{\bar{K}}} = 0.$

1.ii. $\quad \bar{\bar{\gamma}} \geq 0$ *and* $\bar{\bar{\gamma}}\hat{\bar{\bar{K}}} = 0.$

1.iii. *(Euler-Lagrange):*

a) $\dot{\lambda} = -\hat{\mathcal{H}}_x$, *where* $\mathcal{H} = \lambda_0 f^0 + \lambda f + \bar{\mu}\bar{K} + \bar{\bar{\mu}}\bar{\bar{K}} = \mathcal{H}(t, \lambda_0, \lambda, \mu, x, u).$

b) $\hat{\mathcal{H}}_u = 0$

1.iv. $\frac{d}{dt}\hat{\mathcal{H}} = \hat{\mathcal{H}}_t$.

2. *(Weierstrass condition-Pontriagin maximum principle):* (See Pontriagin, etc. (1961)). *With* $\lambda(t)$ *defined in* 1.ii *we have:*

$$\lambda_0 f^0(t, \hat{x}, \hat{u}) + \lambda f(t, \hat{x}, \hat{u}) \geq \lambda_0 f^0(t, \hat{x}, u) + \lambda f(t, \hat{x}, u)$$

for $\hat{x}, u$ *satisfying constraints (1) and (2).*

3. *(Transversality conditions):*

i) $\hat{G}_{t_0} - \hat{\mathcal{H}}|_{t=\hat{t}_0} = 0$

ii) $\hat{G}_{t_1} + \hat{\mathcal{H}}|_{t=\hat{t}_1} = 0$

iii) $\hat{G}_{x_0} + \lambda(\hat{t}_0) = 0$

iv) $\hat{G}_{x_1} - \lambda(\hat{t}_1) = 0$

4. *(Clebsch-Valentine-Berkovitz)* $\pi\hat{\mathcal{H}}_{uu}\pi^* \leq 0$ for π with:

$$\hat{K}^{\alpha}_u \pi^* = 0, \qquad \alpha \in A.$$

5. *(Jacobi-Mayer-Bliss) If, in addition to our assumptions,* Γ *is normal, then:*

6. $$Q(\hat{\Delta}, \lambda_0, \lambda, \mu, \gamma; \delta) = d^2(G(\hat{\Delta}; \tau_0, \tau_1, \xi_0, \xi_1) + [(\hat{\mathcal{H}}_t + \dot{\lambda} f)\tau_r^2 + 2\dot{\lambda}\xi(t_r)\tau_r]\Big|_{r=0}^{r=1}$$

$$+ \int_{t_0}^{\hat{t}_1} [\xi\hat{\mathcal{H}}_{xx}\xi^* + 2\xi\hat{\mathcal{H}}_{xu}v^*]\, dt \leq 0$$ *for* δ *satisfying:*

7.1 $\dot{\xi} = \hat{f}_x\xi^* + \hat{f}_u v^*$

7.2 $\hat{K}^{\alpha}_x\xi + \hat{K}^{\alpha}_u v = 0, \ \alpha \in A$

7.3 $(\hat{g}^{\beta}_{t_0} + \hat{g}^{\beta}_{x_0}\hat{f}|_{t=t_1})\tau_0 + (\hat{g}^{\beta}_{t_1} + \hat{g}^{\beta}_{x_1}\hat{f}|_{t=t_1})\tau_1 + \hat{g}^{\beta}_{t_0}\xi_0 + \hat{g}^{\beta}_{x_1}\xi_1 = 0, \qquad \beta \in B.$

Outline of the Proof:

Introduce the variables s as follows.

8. $\dot{s} = u.$

We may rewrite our problem as:

$$\text{Maximize } J = g(t_0, t_1, x_0, x_1, \int_{t_0}^{t_1} f^0(t, x, \dot{s})\, s \cdot t_0$$

(1)' $\quad f(t, x, \dot{s}) - \dot{x} = 0$

(2.1)' $\quad K^{\overline{\alpha}}(t, x, \dot{s}) = 0,$

(2.2)' $\quad K^{\overline{\overline{\alpha}}}(t, x, \dot{s}) \geq 0,$

and to (3.1) and (3.2).

This is a problem of type A". $\hat{\Gamma} = (\hat{t}_0, \hat{t}_1, \hat{x}_0, \hat{x}_0, \hat{y}_0, \hat{y}_1, \hat{x}, \hat{s})$ is a solution to this problem. Define $F^2 = \lambda_0 f^0 + \lambda(f - \dot{x}) + \overline{\mu}\overline{K} + \overline{\overline{\mu}}\overline{\overline{K}}$, and $G^2 = \lambda_0 g^0 + \overline{\gamma g} + \overline{\overline{\gamma g}}$. Since g^0, g is independent of s_0 and s_1 we have G^2 is identical to G and we obtain condition 1.ii as an immediate result of 1.2 of theorem 1 of chapter 6. That theorem applies as could easily be verified. We further note the $F^2 = \mathcal{H} - \lambda\dot{x}$, and that results in 1.i which follows from 1.1 of theorem 1 of chapter 6, by taking $(\lambda, \overline{\mu})\ \overline{\lambda}$ and $\overline{\overline{\mu}}$ to $\overline{\overline{\lambda}}$. Now by the Euler-Lagrange equations for problem A" we have: $\frac{d}{dt}\hat{F}^2_{\dot{x}} = \hat{F}^2_x$ and $\frac{d}{dt}\hat{F}^2_{\dot{s}} = \hat{F}^2_s$. But $\hat{F}^2_{\dot{x}} = -\lambda$ and $\hat{F}^2_x = \hat{\mathcal{H}}_x$. Thus $\dot{\lambda} = -\hat{H}_x$ which is 1.iii-a of the present theorem. Also $\hat{F}^2_{\dot{s}} = \hat{\mathcal{H}}_{\dot{s}} = \hat{\mathcal{H}}_u$ and $\hat{F}_s = 0$. Thus $\hat{\mathcal{H}}_u = 0$ which is our condition 1.iii-b. $\hat{F}^2_t = \hat{\mathcal{H}}_t$, by definition of F^2 and $\mathcal{H}$. Also $\hat{F}^2_{\dot{s}} = \hat{F}^2_u = 0$ as we have just seen and $\hat{F}^2_{\dot{x}} = -\lambda$. Thus, letting y in problem A" equal (x, s) we have from conditions (1.4) of theorem 1, chapter 6:

$$\frac{d}{dt}(\hat{F}^2 - \dot{\hat{y}}\hat{F}^2_{\dot{y}}) = \frac{d}{dt}(\hat{F}^2 + \lambda\dot{x}) = \frac{d}{dt}\hat{\mathcal{H}} = \hat{F}^2_t = \hat{\mathcal{H}}_t,$$

which is our condition 1.iv.

To prove 2, we write the Weierstrass E function, E^2, for the present problem. Let $p = (p_1, p_2)$ be the "test function" for our problem, corresponding to $(\dot{\hat{x}}, \dot{y}) = (\dot{\hat{x}}, \hat{u})$. Since the constraints must be satisfied for the inequality, 2 of theorem 1, chapter 6, to hold we have $p_1 = f(t, x, u)$. Writing the inequality and noting the relation between F^2 and $\mathcal{H}$ we have:

$$E^2 = \mathcal{H}(t, \hat{x}, y, \lambda_0, \lambda, u) - \lambda f(t, \hat{x}, y) - \mathcal{H}(t, \hat{x}, \hat{u}, \lambda_0, \lambda, \mu) +$$

$$+ \lambda f(t, \hat{x}, \hat{u}) - (f(t, \hat{x}, u) - \hat{\dot{x}})\hat{F}^2_{\dot{x}} - (y - \hat{u})\hat{F}^2_u \leq \bar{\bar{\mu}}\,\bar{K}.$$

But $\hat{F}^2_x = -\lambda$ and $\hat{F}^2_u = 0$. Thus we have:

$$\mathcal{H}(t, \hat{x}, u, \lambda_0, \lambda, \mu) - \mathcal{H}(t, \hat{x}, \hat{u}, \lambda_0, \lambda, \mu) - \lambda f(t, \hat{x}, u) +$$

$$+ \lambda f(t, \hat{x}, \hat{u}) - f(t, \hat{x}, u) + f(t, \hat{x}, \hat{u}) - \bar{\bar{\mu}}\,\bar{K} \leq 0,$$

But $\mathcal{H}(t, \hat{x}, \hat{u}, \lambda_0, \lambda, \mu) = \lambda_0 f^0(t, \hat{x}, \hat{u}) + \lambda f(t, \hat{x}, \hat{u})$, by 1.i and since $K(t, \hat{x}, \hat{u}) = 0$. Also $\mathcal{H}(t, \hat{x}, u) - \bar{\bar{\mu}}\,\bar{K} = \lambda_0 f^0(t, \hat{x}, u) + \lambda f(t, \hat{x}, u)$ in case constraints (2.1) are satisfied. Combining this with our last inequality we get 2 of the theorem.

To prove 3, we write the transversality conditions for the present problem recalling that $G^2 = G$ and substituting $F^2 = \mathcal{H} - \lambda\dot{x}$. By (1.5) of theorem 1 of chapter 6 we have:

9. $\hat{G}_{t_0} - \hat{\mathcal{H}}|_{t = t_0} = 0$
10. $\hat{G}_{t_1} + \hat{\mathcal{H}}|_{t = t_1} = 0$
11. $\hat{G}_{x_0} + \lambda(t_0) = 0$
12. $\hat{G}_{x_1} - \lambda(t_1) = 0.$

Condition 3 of the theorem follows from 9-12. Condition 4 follows from 3 of theorem 1, chapter 6, by noting that $\hat{F}^2_{\dot{y}\dot{y}} = 0$ and that $\hat{F}^2_{uu} = \hat{\mathcal{H}}_{uu}$. 5 follows from 4 of theorem 1, chapter 6 by noting that:

i) $d^2G^2 = d^2G$
ii) $F^2_t = \hat{\mathcal{H}}_t$
iii) $\hat{\dot{x}}\hat{F}^2_x = -\dot{\lambda} f(t, \hat{x}, \hat{u}),\ u\hat{F}^2_z = 0$
iv) $\hat{F}^2_{xx} = \hat{\mathcal{H}}_{xx},\quad \hat{F}^2_{zx} = 0,\quad \hat{F}^2_{zz} = 0.$

v) $\hat{F}^2_{x\dot{x}} = 0,\ \hat{F}^2_{xu} = 0,\ \hat{F}^2_{z\dot{x}} = 0,\quad \hat{F}^2_{zu} = 0.$

vi) $\hat{F}^2_{\dot{x}u} = 0,\ \hat{F}^2_{\dot{x}\dot{x}} = 0,\ \hat{F}^2_{u\dot{x}} = 0,\quad \hat{F}^2_{uu} = \hat{\mathcal{H}}_{uu}.$

This completes an outline of the proof of theorem 1.

Theorem 2. *Sufficient conditions (Mangasarian, Pennissi): Suppose:* 1) g^0, g, f^0, f, *and* K *are of class* C^2. 2) $\hat{\Delta}$ *satisfies conditions 1 and 3 of theorem 1 with* $\lambda_0 > 0$. *Then:*

a) *If* f, $\bar{g}$, *and* K *are linear and if* f^0, $\bar{K}$, *and* $\bar{\bar{g}}$ *are concave, then* $\hat{\Delta}$ *is a global solution of the control problem.*

b) *If* $Q < 0$ *for* $\delta \neq 0$ *satisfying equations (7) of this section, where* Q *is defined in 6 of theorem 1, then* $\hat{\Delta}$ *is a weak local solution of the control problem.*

Theorem 2 follows int a straightforward manner from theorem 2 of chapter 6 after performing the necessary transformations as we did in theorem 1 of this section.

Now we state a necessary and sufficient condition for normality in the control problem. However, the condition is hard to verify and, in many applications, one may verify that $\lambda_0 > 0$ by way of contradiction. This, of course, would not be a proof of normality since the multiplier vector may not be unique.

Remark: The rank condition is necessary and sufficient for Δ to be normal.

Definition: The rank condition for optimal control: We say that the rank conditions for optimal control is satisfied if the matrix

$$[(\hat{g}^{\beta}_{t_0} + \hat{g}^{\beta}_{x_0}\hat{\dot{x}}_0)\tau^{\sigma}_0 + (\hat{g}^{\beta}_{t_1} + \hat{g}^{\beta}_{x_1}\dot{x}_1)\tau^{\sigma} + \hat{g}^{\beta}_{x_0}\xi^{\sigma}_0 + \hat{g}^{\beta}_{x_1}\xi^{\sigma}_1], \qquad \beta \in B,$$

where $\tau^{\sigma}_0, \tau^{\sigma}_1, \xi^{\sigma}_0, \xi^{\sigma}_1$ are the end points of a variation $(\xi^{\sigma}, v^{\sigma})$ with respect to a vector of parameters with as many components as the number of elements of B, has rank equal to the number of elements of B for ever such *admissible variation.* An admissible variation is, as before, a variation such that the differentials of equality constraints and effective inequality constraints, with $(\xi^{\sigma}, v^{\sigma})$ as increments, is zero.

Remark: Confusion continues to surround condition 3 of theorem 1 section 1.2 above especially among economists who write books on optimal control. They seem to think that *any* condition is sufficient for normality and when the mistake is pointed out, they say that they have not seen counter examples. Here then are some counter examples. In fact, here is a continuum of counter examples.

Example Class 1 (Pars (1962)): Consider the problem of maximizing

$\int_0^1 f(t, x_1, x_2, u_1, u_2)dt$ subject to: $\dot{x}_1 = u, \dot{x}_2 = v, u - (1+v^2)^{1/2} = 0.$

Note that the "rank condition", condition 3 of theorem 1 is satisfied. And yet $\lambda_0 = 0$ so we have an abnormality of order 1. This is because of rigidity, these only one "admissible curve" and so we can not "vary a little" to compute our first variation. Indeed a maximum does exist.

Example Class 2: Suppose in section 1.1 above that $f(t, x, u) \equiv 0$ and that the constraints (2.1) and (2.2) are of the form $u \geqq a$ where a is a known function. The "rank condition" is satisfied. The problem now is a finite dimensional problem. Any counter example that shows that a constraint qualification is needed for the first order necessary conditions to be normal is valid here.

Note: I have not counted the elements of class 2 of examples but I trust that the cardinality of the union of the two class of examples is at least c.

At the end of this section we will discuss two special problems:

(a) If, in addition to the constraints (1) - (3) we have: $u \geq 0, x_0 \geq 0,$ $x_1 \geq 0$, then the condition 1.iii-b of theorem 1 becomes $\hat{\mathcal{H}}_u \leq 0$, $\hat{\mathcal{H}}_{u_r} \hat{u}_r = 0$, and conditions 3.iii and 3.iv of the theorem become: $\hat{G}_{x_0} \leq \lambda(t_0)$, $\hat{x}_0^i \hat{G}_{\hat{x}_0^i} = \lambda^i(t_0)\hat{x}_0^i$ and $\hat{G}_{x_1} \leq \lambda(t_1)$, $\hat{x}_1 \hat{G}_{x_0}{}^1 \lambda(t_1)$. This could be easily seen by considering the non-negativity constraints as additional inequality constraints of the form (2.2) and (3.2).

(b) Suppose our objective function if of the form:

$$g^0 + W\left[\int_{t_0}^{t_1} C(t, x, u)\, dt\right],$$ where C is an n-vector valued function. Then the Hamiltonian for the problem becomes: $\mathcal{H} = \lambda_0 W_C C + \lambda f$ and G remains unchanged. This could be seen by introducing the variable θ as follows:

15. $\dot{\theta} = C(t, x, u)$, $\theta(t_0)$ and $\theta(t_1)$ free.

The problem then becomes: max. $g^0 + W(\theta_1 - \theta_0)$ subject to 1 - 3 and to 15. The multipliers corresponding to 15 are constants and their values are determined from the transversality conditions of theorem 1. This problem was first formulated by Brady (1941).

2. EXTENSIONS

PROBLEM WITH BOUNDED STATE VARIABLES

If we have, in addition to 1 - 3, constraints of the form:

16. $M^j(t, x) \geq 0$, $j = 1, \ldots, x$,

then the rank condition in the first order necessary theorem could not be satisfied. Here we state a modification to the Euler-Lagrange equations and the Pontriagin maximum principle of that theorem, due to Russak (1970). Add to the assumptions of theorem 1 the assumption: M is of class C^2. Introduce the functions $N(t, x, u) = M_t + M_x f(t, x, u)$ and the multipliers $q(t)$. Define $\mathcal{H}'' = \mathcal{H} + qN$. Then, if $\hat{\Delta}''$ is a solution to this problem, conditions 1, 2, and 3 of the theorem hold for $\mathcal{H}''$ with G unchanged. Furthermore:

(i) $q(t) \geq 0$,

(ii) $\dot{q}(t) \leq 0$,

(iii) $\dot{q}(t) = 0$ of intervals with $M(t, \hat{x}) > 0$ and

(iv) $q(t_r)M(\hat{t}_r, \hat{x}_r) = 0, \qquad r = 0, 1.$

PROBLEM WITH TIME LAGS

We discuss a control problem with time lags in both state and control variables with equality and inequality constraints. We present a conjecture about the Euler-Lagrange equations and the Pontriagin maximum principle. In the absence of inequality constraints, the conjecture was proved by Halanay* (1968). Consider a system given by:

17. $\dot{x}(t) = F(t, x, x^{-1}, \ldots, x^{-k}, u, u^{-1}, \ldots, u^{-k})$,

where $x = x(t)$ is an n-vector valued function, $u = u(t)$ is an R-valued function, $x^{-s} = x(t - \tau^s)$ and $u^{-s} - u(t - \tau^s)$ with $s = 1, \ldots, k$, τ^s fixed and positive. let t_0 and t_1 be fixed and let $x(t)$ be a given function on the interval $[t_0 - \max_\beta \tau^s, t_0]$. Our problem is to maximize the functional:

$$g_0(x_1) + \int_{t_0}^{t_1} f^0(t, x, x^{-1}, \ldots, x^{-k}, u, u^{-1}, \ldots, u^{-k}),$$

subject to:

17.1. $K^{\bar{\alpha}}(t, x, x^{-1}, \ldots, x^{-k}, u, u^{-1}, \ldots, u^{-k}) = 0, \qquad \bar{\alpha} = 1, \ldots, m_1.$

17.2. $K^{\bar{\bar{\alpha}}}(t, x, x^{-1}, \ldots, x^{-k}, u, u^{-1}, \ldots, u^{-k}) \geq 0, \qquad \bar{\bar{\alpha}} = m_1 + 1, \ldots, m.$

17.3. $g^{\bar{\beta}}(x_1), \quad \bar{\beta} = 1, \ldots, p_1.$

17.4. $g^{\bar{\bar{\beta}}}(x_1), \quad \bar{\bar{\beta}} = p_1 + 1, \ldots, p.$

---

* Halanay's description of the control systems is much more general than the one presented here.

Let η be a time function, by η^{-s} we denote $\eta(t - \tau^s)$, by η^r we denote $\eta(t + \tau^r)$ and by η^{r-s} we denote $\eta(t - \tau^s + \tau^r)$. Let $\lambda(t)$, $\bar{\mu}(t)$, and $\bar{\bar{\mu}}(t)$ be time functions with the dimensions f, K, and $\bar{K}$ and let $\lambda_0 > 0$ be a constant scalar. Let $\bar{\gamma}$ and $\bar{\bar{\gamma}}$ be constant vectors with the dimensions of $\bar{g}$ and $\bar{\bar{g}}$. Define:

$$\mathcal{H}^0 = \lambda_0 f^0 + \lambda f + \bar{\mu}K + \bar{\bar{\mu}}\bar{K} = \lambda_0 f^0 + \lambda f + \mu K.$$

$$\begin{aligned}\mathcal{H}^1 &= \lambda_0 f^0(\tau + \tau^1, x^1, x, x^{1-2}, \ldots, x^{1-k}, u^1, u, \ldots, u^{1-k}) + \\ &+ \lambda^1 f(\tau + \tau^1, x^1, x, x^{1-2}, \ldots, x^{1-k}, u^1, u, \ldots, u^{1-k}) + \\ &+ \mu^1 K(\tau + \tau^1, x^1, x, x^{1-2}, \ldots, x^{1-k}, u^1, u, \ldots, u^{1-k}).\end{aligned}$$

$$\begin{aligned}\mathcal{H}^2 &= \lambda_0 f^0(\tau + \tau^2, x^2, x^{2-1}, x, \ldots, x^{2-k}, u, u^{2-1}, u, \ldots, u^{2-k}) + \\ &+ \lambda^1 f(\tau + \tau^2, x^2, x^{2-1}, x, \ldots, x^{2-k}, u, u^{2-1}, u, \ldots, u^{2-k}) + \\ &+ \mu^1 K(\tau + \tau^2, x^2, x^{2-1}, x, \ldots, x^{2-k}, u, u^{2-1}, u, \ldots, u^{2-k}).\end{aligned}$$

$$\begin{aligned}.\mathcal{H}^k &= \lambda_0 f^0(\tau + \tau^k, x^k, x^{k-1}, \ldots, x, u, u^{k-1}, \ldots, u) + \\ &+ \lambda^1 f(\tau + \tau^k, x^k, x^{k-1}, \ldots, x, u, u^{k-1}, \ldots, u) + \\ &+ \mu^1 K(\tau + \tau^k, x^k, x^{k-1}, \ldots, x, u, u^{k-1}, \ldots, u).\end{aligned}$$

$$\mathcal{H} = \sum_{r=0}^{k} \mathcal{H}^r.$$

$$G = \lambda_0 g^0 + \bar{\gamma}\bar{g} + \bar{\bar{\gamma}}\bar{\bar{g}}.$$

Now we state our conjecture:

Assume: 1) f^0, f, K *and* g *are of class* C^2, 2) $\hat{\Delta} = (\hat{u}, \hat{x}, \hat{x}_1)$ *is a solution to our problem. Then there exist constants* (λ_0, γ) *and functions* (λ, μ) *with*

1.i) $\bar{\bar{\mu}} \geq 0$, $\bar{\bar{\mu}}\hat{\bar{K}} = 0$

1.ii) $\bar{\bar{\gamma}} \geq 0$, $\bar{\bar{\gamma}}\bar{\bar{g}} = 0$

2) *Euler-Lagrange-Halanay equations:*

2.i) $\dot{\lambda} = - \hat{\mathcal{H}}_x$

2.ii) $\hat{\mathcal{H}}_u = 0$

3) *Pontriagin maximum principle:*
Setting $x, x^{r-s}, x^s = (\hat{x}, \hat{x}^{r-s}, \hat{x}^s)$ *and* $u^{r-s}, u^s = \hat{u}^{r-s}, \hat{u}^s$ *for* $s \neq r$, $s \neq 0$ *in* $\mathcal{H}$, $\mathcal{H}$ *is maximized in* u *at* $\hat{u}$.

4) *Transversality:*

$$\hat{G}_{x_1} - \lambda(t_1) = 0, \quad \lambda(t_1 + \theta) = 0, \quad \textit{for } \theta > 0.$$

A CONTROL PROBLEM WITH A VECTOR CRITERION

For the purpose of this paragraph let g^0 and f^0 be N-vector functions. The problem is to maximize, in the sense of Pareto, the vector of functionals:

$$J[u] = g^0(x_0, x_1) + \int_{t_0}^{t_1} f^0(t, x, u)\, dt,$$

subject to constraints 1 - 3. Arguing exactly as we did in the finite dimensional problem, a solution, Δ, to this problem is a solution to the following N problems:

Problem $A_{\underline{i}}$: Maximize $g^{0^{\underline{i}}} + \int_{t_0}^{t_1} f^{0^{\underline{i}}}(t, x, u)\, dt$, subject to:

$$19. \quad g^{0^i}(x_0, x_1) + \int_{t_0}^{t_1} f^{0^i}(t, x, u)\, dt \geq \hat{g}^{0^i}(\hat{x}_0, \hat{x}_1) + \int_{t_0}^{t_1} \hat{f}^{0^i}(t, \hat{x}, \hat{u})\, dt, \ i \neq \underline{i}.$$

For each problem $A_{\underline{i}}$ we get a vector of multipliers $(\lambda_0^{\underline{i}}, \gamma^{\underline{i}}) \neq 0$, $(\lambda^{\underline{i}}, \gamma^{\underline{i}})$ with $(\lambda_0, \lambda^{\underline{i}}, \gamma^{\underline{i}}) \neq 0$, where $\lambda^{\underline{i}}$ is a non-negative N-vector such that conditions 1 - 3 of theorem 1 of this chapter are satisfied. Summing each of these conditions over $\underline{i}$ from 1 to N and defining $\lambda_0 = \sum_{\underline{i}} \lambda_0^{\underline{i}}, \gamma = \sum_{\underline{i}} \gamma^{\underline{i}}, \gamma = \sum_{\underline{i}} \lambda^{\underline{i}}, \mu = \sum_{\underline{i}} \mu^{\underline{i}}$, we may show that: *There exists a vector of non-negative constants* λ_0, *a vector of constants* γ, *and a vector function* (λ, μ) *such that conditions 1 - 3 of theorem 1 of this chapter hold, with* $\lambda_0 f^0$ *and* $\lambda_0 g^0$ *interpreted as dot products.*

SENSITIVITY ANALYSIS

In this section we use a technique of proof that was used by Mangasarian (1966) to derive some inequalities concerning the effect of a change in the exogenous variables on the solutions of an optimal control problem. The solution of our control problem depends on the values of some exogenous variables, i.e., variables that are not determined by the solution of the problem. Once these variables are given, the problem may be solved in the usual ways. Rather than solve for the optimal controls and state variables, it might be of interest to obtain pairwise comparisons of optimal decisions under pairs of "values" of the exogenous variables.

Let a system be given by

$$\dot{x}_t = f(t, x_t, u_t, \beta_t, B) \tag{1}$$

where $x \in E^{n1}$, $u \in E^{n2}$, $\beta \in E^{n3}$, $B \in E^{n4}$, and where β_t and B are not choice variables. Suppose the control constraints are given by

$$h^{\alpha}(t, x_t, u_t, \beta_t, B) \tag{2}$$

Let the endpoint conditions be given by

$$g^{\delta}(x_{t_0}, x_{t_1}, B) \geq 0, \qquad \delta = 1, ..., \bar{\delta} \tag{3}$$

The objective of the system is to maximize

$$W(I, x_{t_0}, x_{t_1}, B) \tag{4}$$

where

$$I = \int_{t_0}^{t_1} \phi(t, x_t, u_t, \beta_t, B).$$

Assume the functions and admissible controls satisfy all of the regularity conditions that assure that the first order necessary condition theorems apply, e.g., those given in Hestenes (1965). Let (β_t^1, B^1) and (β_t^2, B^2) be two values of the exogenous variables. Then we would have two, not necessarily distinct, solutions $z^1 = x_t^1, u_t^1, x_{t_0}^1, x_{t_1}^1$ and $z^2 = x_t^2, u_t^2, x_{t_0}^2, x_{t_1}^1$ of the control problem. We use the superscript ij on a function to indicate that it is evaluated at the value i of the variables z and the value j for the variables (β_t, B), i, j = 1, 2. We use the subscripts z, β_t, B to indicate differentiation with respect to these variables. The first order conditions optimal solutions are the existence of time functions $\lambda^i(t)$ and $\mu^i(t)$ and of constants λ_0^i, γ^i, where $\lambda^i(t)$ has values in E^n, $\mu^i(t)$ has values in E^{α}, $\lambda_0^i \geq 0$ is a scalar and $\gamma^i \in E^{\alpha}$, such that:

$$\dot{\lambda}^i = - H_x^{ii}, \qquad i = 1, 2 \tag{5}$$

where

$$H = \lambda_0 w_I \phi + \lambda f + \mu h$$

$$\mu^i \geq 0, \mu^i h^{ii} = 0, \qquad i = 1, 2 \tag{6}$$

$$\gamma^i \geq 0, \gamma^i g^i = 0, \qquad i = 1, 2 \tag{7}$$

$$H_u^{ii} = 0, \qquad i = 1, 2 \tag{8}$$

$$\lambda^i(t_0) + G_{x_{t_0}}^{ii} = 0, \qquad i = 1, 2$$

where

$$G = \lambda_0 w + \gamma g \tag{9}$$

$$\lambda^i(t_i) - G^{ii}_{x_{t_1}} = 0. \tag{10}$$

Assume that w, f, h, *and* g *are concave in* z *for fixed* (β_t, B), *are differentiable and that either* $\lambda^i \geq 0, i = 1, 2$, *or* f *is linear.* Then the above conditions are also sufficient, if $\lambda_0 > 0$, as could be easily seen by using Mangasarian's methods (1966). From the concavity of w it follows that

$$\begin{aligned} w^{22} - w^{11} &= w^{22} - w^{21} + w^{21} - w^{11} \\ &\leq w^{22} - w^{21} + w^{11}_I \Delta I + w_{x_{t_0}} \Delta x_{t_0} + w_{x_{t_1}} \Delta x_{t_1} \end{aligned} \tag{11'}$$

where $\Delta I = \int_{t_0}^{t_1} (\phi^{21} - \phi^{11})dt$ and $\Delta z = z^2 - z^1$. Relation (11'), assuming $w_t \geq 0$, may be written in view of the concavity of ϕ as

$$\begin{aligned} w^{22} - w^{11} \leq w^{22} - w^{21} + w^{11}_I \left(\int_{t_0}^{t_1} [\phi^{11}_x \Delta x + \phi^{11}_u \Delta u]\, dt \right) \\ + w_{x_{t_0}} \Delta x_{t_0} + w_{x_{t_1}} \Delta x_{t_1}. \end{aligned} \tag{11}$$

We also have, by assumptions about λ^1 and f,

$$\lambda^1 \Delta \dot{x} = \lambda^1 (f^{22} - f^{11}) \leq \lambda^1 (f^{22} - f^{21}) + \lambda^1 f^{11}_u \Delta x + \lambda^1 f^{11}_u \Delta u. \tag{12}$$

By assumptions about h and g we have, using (6) and (7),

$$0 \leq \mu^1 (h^{22} - h^{11}) = \mu^1 (h^{22} - h^{21}) + \mu^1 h^{11}_x + \mu^1 h^{11}_u \tag{13}$$

$$0 \leq \gamma^1 (g^{22} - g^{11}) = \gamma^1 (g^{22} - g^{21}) + \gamma^1 g^{11}_{x_{t_0}} \Delta x_{t_0} + \gamma^1 g^{11}_{x_{t_1}} \Delta x_{t_1}. \tag{14}$$

Multiplying both sides of (11) by λ_0, integrating both sides of (12) and adding the results to (13) and (14) we get, in view of (5), (8), (9), and (10)

$$\lambda_0^1(w^{22} - w^{11}) + \int_{t_0}^{t_1} \lambda^1 \Delta\dot{x} \le \lambda_0^1(w^{22} - w^{21}) + \gamma^1(g^{22} - g^{21}) + \lambda_0^1(t_1)\Delta x_{t_1}$$

$$- \lambda^1(t_0)\Delta x_{t_0} - \int_{t_0}^{t_1} \dot{\lambda}^1 \Delta x \, dt + \int_{t_0}^{t_1} [\lambda^1(f^{22} - f^{21}) + \mu^1(h^{22} - h^{21})] \, dt. \qquad (15)$$

Thus, since $\int_{t_0}^{t_1} (\lambda^1 \Delta\dot{x} + \dot{\lambda}^1 \Delta x) \, dt = \lambda^1(t_1)\Delta x_{t_1} - \lambda^t(t_0)\Delta x_{t_0}$, we have

$$\lambda_0^1(w^{22} - w^{11}) \le \lambda_0^1(w^{22} - w^{21}) + \gamma^1(g^{22} - g^{21})$$

$$+ \int_{t_0}^{t_1} [\lambda^1(f^{22} - f^{21}) + \mu^1(h^{22} - h^{21})] \, dt. \qquad (16)$$

Assuming concavity in (β_1, B) for any fixed z and denoting $\beta_t^2 - \beta_t^1$ by $\Delta\beta$ and $B^2 - B^1$ by ΔB we have

$$\lambda_0^1(w^{22} - w^{11}) \le \int_{t_0}^{t_1} \{[\lambda_0^1 w_I^{21} \phi^{22} + \lambda^1 f_\beta^{21} + \mu^1 h_\beta^{21}]\beta\Delta$$

$$+ [\lambda_0^1 w_I^{21} \phi_B^{21} + \lambda^1 f_B^{21} + \mu^1 h_B^{21}]\Delta B\} \, dt + (\lambda_0^1 w_B^{21} + \gamma^1 g_B^{21})\Delta B \qquad (17)$$

Repeating the process leading to (16) we have

$$\lambda_0^2(w^{11} - w^{22}) \le \lambda_0^2(w^{11} - w^{12}) + \gamma^2(g^{11} - g^{12})$$

$$+ \int_{t_0}^{t_1} [\lambda^2(f^{11} - f^{12}) + \mu^2(h^{11} - h^{12})] \, dt. \qquad (18)$$

Assuming concavity in β_t and B for any fixed z we have

$$\lambda_0^2(w^{11} - w^{22}) \le \int_{t_0}^{t_1} \{[\lambda_0^2 w_I^{12} \phi_\beta^{22} + \lambda^2 f_\beta^{12} + \mu^2 h_\beta^{12}](-\Delta\beta)$$

$$+ [\lambda_0^2 w_I^{12} \phi_B^{12} + \lambda^2 f_B^{12} + \mu^2 h_B^{12}](-\Delta B)\} \, dt + (\lambda_0^2 w_B^{12} + \gamma^2 g_B^{12})(-\Delta B). \qquad (19)$$

The inequalities (16) - (19) were derived without assuming λ_0^i to be nonzero. If *we assume that* $\lambda_0^i > 0$ then they may be set equal to unity and we have some further results. Adding (16) and (18) we have

$$(w^{22} - w^{21}) + (w^{11} - w^{12}) + \gamma^1(g^{22} - g^{21}) + \gamma^2(g^{11} - g^{12})$$

$$+ \int_{t_0}^{t_1} [\lambda^1(f^{22} - f^{21})$$

$$+ \lambda^2(f^{11} - f^{12}) + \mu^1(h^{22} - h^{21}) + \mu^2(h^{11} - h^{12})]\, dt \geq 0. \tag{20}$$

We also get from (17) and (19)

$$\int_{t_0}^{t_1} \{[(w_I^{21}\phi_\beta^{21} - w_I^{12}\phi_\beta^{12}) + (\lambda^1 f_\beta^{21} - \lambda^2 f_\beta^{12}) + (\lambda^1 h_\beta^{21} - \lambda^2 h_\beta^{12})]\Delta B$$

$$+ [(w_I^{21}\phi_B^{21} - w_2^1\phi_B^{12}) + (\lambda^1 f_B^{21} - \lambda^2 g_B^{12})\Delta B + (\lambda^1 h_B^{21} - \lambda^2 h_B^{12})]\Delta B\}$$

dt

$$+ [(w_B^{21} - w_B^{12}) + (\gamma^1 g_B^{21} - \gamma^2 g_B^{12})]\Delta B \geq 0. \tag{21}$$

In conclusion we mention that these results may be used in numerous applications. For instance, in national economic planning they may be used to study the effects of changes in resource endowments and in the study of the effects of changes in population growth patterns. Relations (16)-(19) may be used to determine the effect on the maximand of a change in the exogenous variables. Relations (20) and (21) may be used to determine the effect on the state and control variables of a change in the exogenous variables.

3. TWO STATIC ECONOMIC EXAMPLES

The vast majority of applications of calculus of variations to economic analysis have been dynamic, in the sense of choosing among time paths. This was done by using criterion functionals that are integrals of instantaneous criterion functions, which involves a sever restriction on the form of such functionals.

We shall present, where, two examples from economic analysis where the use of integrals is more natural. The problems are of some interest in themselves. The first problem deals with Ricardian rent and was first solved by Samuelson (1959). We shall provide a reformulation and proof of the results of the appendix of (1959).

The second problem attempts to answer a question raised by Qayum (1963) about investment maximizing tax schedules; whether they are progressive or regressive. We answer the question for the case considered by Qayum (1963). In general the answer depends, even in our very simple model, on such factors as the minimum amount of taxes to be collected and on the nature of the frequency distribution of income.

RICARDIAN RENT

We shall use Samuelson's notation and formulation of the problem in Samuelson (1959). The problem of determining the margin of expansion may be stated as follows:

$$\text{Maximize} \quad 2\Pi \int_0^R f^1[x^0(\rho)L^1(\rho), L^1(\rho); \rho]\rho \, d\rho$$

Subject to:

1) $$x - 2\Pi \int_0^R x^1(\rho)\rho \, d\rho \geq 0,$$

2) $$\Pi R^2 - 2\Pi \int_0^R L^1(\rho)\rho \, d\rho \geq 0,$$

3) $$L^1(\rho) \geq 0, \quad x^1(\rho) \geq 0.$$

Proposition: If f is continuously differentiable then a necessary conditions for $\hat{x}^1(\rho)$, $\hat{L}^1(\rho)$, and $\hat{R} > 0$ to be a solution to our problem is that there exist a land rental and a labor wage; $P_L > 0$ and $P_x > 0$ such that,

1) Marginal productivity of labor $\leq P_x$ with equality if $x^1(\rho)$ is positive, i.e.

(i) $\hat{f}_1\hat{L}^1(\rho) \leq P_x$, with equality if $\hat{x}(\rho) > 0$, for $0 < \rho \leq \hat{R}$.

2) Marginal productivity of land $\leq P_L$ with equality if $\hat{x}^1(\rho) > 0$, i.e.

(ii) $\hat{f}_1\hat{x}^1(\rho) + \hat{f}_2 \leq P_L$, with equality if $\hat{L}^1(\rho) > 0$, for $0 < \rho \leq \hat{R}$.

3) At the optimal frontier, the "rent" is at most zero, i.e.

(iii) $f^1[\hat{x}^1(\hat{R})\hat{L}(\hat{R}), \hat{L}^1(\hat{R}); R] - P_x\hat{X}(R) - P_L\hat{L}(\hat{R}) \leq 0$, where rent is defined as the left hand side of (iii).

(iv) $P_x(x - 2\Pi \int_0^{\hat{R}} \hat{x}^1(\rho)\rho \, d\rho) = 0, \; x - 2\Pi \int_0^{\hat{R}} \hat{x}^1(\rho)\rho \, d\rho \geq 0.$

(v) $P_L(\Pi\hat{R}^2 - 2\Pi \int_0^{\hat{R}} \hat{L}(\rho)\rho \, d\rho) = 0, \; \Pi\hat{R}^2 - 2\Pi \int_0^{\hat{R}} \hat{L}(\rho)\rho \, d\rho \geq 0.$

Proof: The proposition is established by applying the theorem of after some transformations - theorem of .[1]

The necessary conditions are: there exist constants $\lambda_0, \lambda_1, \lambda_2$, non-negative and not all zero such that:

(1) $2\Pi\lambda_0 \hat{f}_1\hat{L}\rho - 2\Pi\lambda_1\rho \leq 0$, with equality is $\hat{x}(\rho) > 0$; $\rho \in [0,\hat{R}]$.

(2) $2\Pi\lambda_0 (\hat{f}_1\hat{x} + \hat{f}_2)\rho - 2\Pi\lambda_2\rho \leq 0$, with equality if $\hat{L}(\rho) > 0$; $\rho \in [0,\hat{R}]$.

(3) $-[2\lambda_0\Pi\hat{R}\hat{f}[\hat{x}^1(\hat{R})\hat{L}^1(\hat{R}),\hat{L}^1(\hat{R}); \hat{R}] - 2\lambda_1\pi\hat{x}(\hat{R})\hat{R} - 2\lambda_2\pi\hat{L}(\hat{R})\hat{R}] - 2\Pi\lambda_2\hat{R} = 0.$

---

[1] Samuelson, in (1959), cites Bliss (1946) where there is no treatment of problems with inequality constraints.

$$(4)\ \lambda_1(x - 2\Pi \int_0^{\hat{R}} \hat{x}^1(P)\rho\, d\rho) = 0.$$

$$(5)\ \lambda_2(\Pi\hat{R}^2 - 2\Pi \int_0^{\hat{R}} \hat{L}(\rho)\rho\, d\rho) = 0,$$

Clearly $\lambda_0 > 0$; for if it is zero then so would be λ_1 and λ_2 (by (1) and (2)) which would be a contradiction. Thus we may take $\lambda_0 = 1$. Take $P_x = \lambda_1$ and $P_1 = \lambda_2$. Then (i) and (ii) follow from (1) and (2) for $\rho > 0$. And (iv) and (v) follow from (4) and (5) and the fact that $\hat{x}^1$ and $\hat{L}^1$ satisfy the constraints.

(iii) follows from (3) since we have (by(3)):

$$\hat{f} - P_x\hat{x}(\hat{R}) - P_L\hat{L}(\hat{R}) = -P_L \leq 0.$$

INVESTMENT MAXIMIZING TAX POLICY[2]

In (1963) Qayum asks the following question: Assuming that investment depends only on the rate of income taxation, is an investment maximizing tax policy regressive or progressive? To answer the question Qayum restricts himself to the class of quadratic, in income, tax rates. Then he determines the parameters that lead to selecting one function from this class by finite dimensional maximization methods. However, since he takes income as a continuum, the problem of choosing the optimal tax schedule is a problem of selecting a function from an admissible class of functions defined on the range of taxable income, i.e., we must be able to choose the optimal form and parameters of the function. In fact, we shall show that if an "optimal"[3] tax schedule exists, for the Qayum problem, among the class of differentiable schedules then we get a completely different form of the schedule.

We formulate the problem as a simple problem in the calculus of variations and characterize the solution of the general problem. We

---

[2]I am grateful to Professor Qayum for suggesting this problem.

[3]In the sense of investment maximization.

then solve the problem where the investment function and the frequency distribution of income are those assumed by Qayum.

Let y denote a certain level of income, and suppose y_0 and y_1 are the lowest and highest levels of taxable income. Denote by Y, the interval $[y_0, y_1]$. Let $f(y)$ be the frequency distribution of income. f is defined and is assumed to be differentiable and monotone decreasing on Y. Let $R(y)$ denote the amount of taxing income at level y. The function $R(y)$ will be the unknown in the present problem. We shall pick it from among the class of piecewise differentiable[4] functions defined on Y with values between zero and one, i.e.

(1) $R(t) \in [0, 1]$.

Let $t(y)$ be the tax paid by an individual with income y. Then $t(y) = \int_{y_0}^{y} R(\eta)\, d\eta$. We represent this by way of the following differential equation:

(2) $\dot{t}(y) = R(y), \qquad t(y_0) = 0,$

where the dot over a variable will always represent differentiation with respect to y. The taxes collected from all people with income y is $f(y)t(y)$ and total tax collection, T, is given by the functional:

(3) $T[R] = \int_{y_0}^{y_1} t(y)f(y)\, dy.$

Let $S(r(y))$ denote the proportion that is invested out of an income y. Then investment out of an income y is $yS(R)$ and the investment of all persons with given by $yS(R)f(y)$. Thus total investment, I, is given by the functional:

(4) $I = I[R] + \omega = \int_{y_0}^{y_1} yS(R)f(y)\, dy + \omega.$

---

[4]The choice of this class is dictated by pedagogical considerations.

where the constant ω represents investment out of non-taxable income.

The problem of maximizing investment subject to a lower bound, say γ, on tax collection is equivalent to maximizing $I[R]$, since ω is a constant, subject to constraint (1), (2) and

(5) $T[R] - \gamma \geq 0.$

The following conditions are necessary[5] and sufficient[6] for R to be solution, assuming that S is a concave function of R and that the second derivative of S exists and is continuous.

There exist constants λ_0 and β and functions $\lambda(y)$, $\mu_1(y)$, and $\mu_2(y)$ such that

(6.1) $\lambda_0 \geq 0, \quad \beta \geq 0, \qquad \beta(T[\hat{R}] - \gamma) = 0.$

(6.2) $\dot{\lambda} = -\beta f(y), \qquad \lambda(y_1) = 0.$

(6.3) $\mu_1 \geq 0, \quad \mu_1(1 - \hat{R}) = 0.$

(6.4) $\mu_2 \geq 0, \quad \mu_2\hat{R} = 0.$

(6.5) The vector $(\lambda_0, \lambda, \mu_1, \mu_2, \beta)$ is never a zero vector.

(6.6) $\lambda_0 y f S'(\hat{R} + \lambda(t) - \mu_1(t) + \mu_2(t) = 0.$

From (6.2) it follows that:

(7) $\lambda(y) = \beta \int_y^{y_1} f(\eta)\, d\eta,$

implying that $\lambda(y) \geq 0$ on Y.

(A) $\hat{R}(y) < 1$ for some subinterval of Y with positive length. We can then show that $\lambda_0 > 0$ and hence may be taken equal to 1. This follows, by contradiction, from (6.6), (7), and (6.5). For if $\lambda_0 = 0$ then, since $\mu_1 = 0$, $\lambda(y) = -\mu_2 \leq 0$ and hence $\lambda_2 = 0$. This by (7), implies $\beta = 0$ contradicting (6.5). Conditions (6.6) may now be written:

---

[5]See theorem 1 of this chapter.

[6]See theorem 2 of this chapter.

(8) $yfS'(\hat{R}) = \mu_1(t) - \lambda(t) - \mu_2(t)$.

Assuming that $\hat{R}$ is in (0, 1) and that $S' < 0$ and $S'' < 0$ it is necessary and sufficient for it to be a solution that:

(9.1) $-yfS'(\hat{R}) = \lambda(y)$.

(9.2) $\dot{\lambda}(y) = -\beta f$.

Differentiating both sides of (9.1) with respect to y we have:

(9.3) $\dot{\lambda} = -(fS' + y\dot{f}S' + yfS''\dot{R})$.

By (9.2) and (9.3) we have:

(10) $\dot{R}yS'' = \beta - S'(1 + y\dot{f}/f)$.

From (10) we see that, since $S'' \leq 0$, the sign of $\dot{R}$ is opposite to the sign of $\beta - S'(1 + y\dot{f}/f)$. In case $(1 + y\dot{f}/f) > 0$ the tax is regressive. This, however, occurs when either $\dot{f}/f$ is positive, which does not make sense[7], or when $\dot{f}/f < 0$ and $y|\dot{f}/f| < 1$. If $(1 + y\dot{f}/f) < 0$ then the sign of $\dot{R}$ depends on the magnitude of β in relation to $|1 + y\dot{f}/f|$.

By way of illustration we analyze the following example, considered by Qayum. Let $f = Cy^{-m}$, $m > 2$, $S = AR - BR^2$, $A \leq 2B$. Note that S is concave, and that our analyses apply. By (7), $\lambda = \beta \int_y^{y_1} f(\eta)\, d\eta = C\beta \int_y^{y_1} \eta^{-m}$, i.e.

(11) $$\lambda = \frac{C\beta}{1 - m}(y_1^{1-m} - y^{1-m}).$$

Assuming that $0 < \hat{R} < 1$, (9.1) applies and, by (11),

---

[7] And does not happen anyway.

$$(12) \qquad R = \frac{A}{2B} + \frac{\beta}{2(1 - m)B}\left(\left(\frac{y}{y_1}\right)^{1-m} - 1\right).$$

Differentiating with respect to y, we find that $\dot{R} = \frac{(m - 1)\beta}{2(1 - m)B}\frac{y^{m-2}}{y_1^{m-1}} = -\frac{\beta}{2B}\frac{y^{m-2}}{y_1^{m-1}} \leq 0$. Thus the optimal tax schedule is regressive for all levels of income, contrary to Qayum's assertion that there are cases where the tax is progressive. We may check this last result by (10); substituting for f, S' and S'' we have:

$$(13) \qquad \dot{R}y(-2B) = \beta - (A - 2BR)(1 - m), \text{ i.e.}$$

$$\dot{R} = \frac{-\beta}{2By} + \frac{(A - 2BR)}{2By}(1 - m)$$

$$\leq \frac{-\beta}{2By} + \frac{2B(1 - R)}{2By}(1 - m)$$

$$= \frac{-\beta}{2By} - (1 - R)(1 - m) < 0, \text{ since } m > 2 \text{ and } R < 1.$$

4. INVESTMENT DEMAND

It can be easily shown that a log-linear technology induces, under the assumption of profit maximization by firms, a Leontief structure in value terms in the static case.

We show here that if firms maximize present values of net cash follows, then the interest payments on the value of output capital stocks are indeed proportional to the value of output. The factor of proportionality is, as with the other inputs, constant as long as the technology remains unchanged. The interest payments on the value of the capital stock of the firm can be interpreted as the interest paid on the perpetual bond issued by the firm to finance the acquisition of the capital stock, in this sense it is the actual cost of the input capital.

It is enough to consider only a given sector in the economy. We assume only one firm in this sector, say be taking a representative firm and blowing it up. We identify our firm with our sector; sector j. Let $X_j(t)$ be the output of sector j and let $X_{ij}(t)$ be, noncapital, input i used in sector j all at time t, i = 1,..., n. Let $K_j(t)$ be the stock of capital in sector j at time t. The inputs and output are related according to the production function:

$$(1)\quad \log X_j(t) = b(t) + \sum_i a_{ij}(t)\log X_{ij}(t) + c_j(t)\log K_j(t).$$

And, as usual, we have:

$$(2)\quad \frac{d}{dt}K_j(t) = I_j(t) - \alpha_j(t)K_j(t),$$

where I_j is gross investment by the firm and α_j is the rate of depreciation.

The firm maximizes the present value of its net cash flow i.e. it maximizes:

$$(3)\quad \int_0^\infty [P_j(t)x_j(t) - \sum q_i(t)X_{ij}(t) - \rho(t)I_j(t)]\, e^{-R(t)}dt,$$

where $P_j(t)$ and $q_i(t)$ are prices of output j and inputs i at time t and where $\rho_j(t)$ is the price of new capital goods and $R(t) = \int_0^\infty r(s)\, ds$ with $r(t)$ denoting the discount rate at time t.

Substituting from (1) and (2) into (3) we obtain a simple calculus of variations problem. Assuming a solution to exist and to be non-corner, the Euler equation yields (for all t):

$$(4)\quad a_{ij}(t)P_j(t)X_j(t) = q_i(t)X_{ij}(t)$$

$$(*)\quad \frac{d}{dt}(-\rho_j(t)e^{-R(t)}) = [(c_j(t)P_j(t)X_j(t)/K_j(t) - \rho(t)\alpha_j(t)]e^{-R(t)}$$

From (*) we get

$$(5)\quad [r(t) - \dot{\rho}(t)/\rho(t) + \alpha_j(t)]\rho(t)K_j(t) = cj(t)P_j(t)X_j(t).$$

Let us use lower case letters to denote things being measured in value terms.

Then (4) and (5) yields a Leontief structure for input demand in value terms:

$$(6) \qquad x_{ij}(t)/x_j(t) = a_{ij}(t)$$

$$(7) \qquad \delta_j(t)k_j(t)/x_j(t) = c_j(t)$$

where $\delta(t) = (r(t) + \alpha_j(t) - \dot{\rho}(t)/\rho(t))$ is the interest rate adjusted for the change in prices and for depreciation rate.

From (7) we can derive the form of the conditional investment demand function. It has the form:

$$(8) \quad I_j = \backslash f(d,dt)\,(c_jx_j/\rho\delta_j) + (\alpha_j/\rho_j\delta_j)k_j = \backslash f(c_jx_j,\rho_j\delta_j)\ \backslash b(\backslash f(\backslash o(c,\dot{})j,cj)$$

$$+ \frac{\dot{x}j}{xj} - \frac{\dot{\delta}j}{\delta j} - \frac{\dot{\rho}j}{\rho j} + \alpha_j)$$

Thus the demand for investment by a profit maximizing firm depends on prices and interest rates as well as the rates of change in these variables. It depends also on the acceleration of price of capital goods and on the rate of growth of output and represents the first theoretical justification for the dependence of demand on prices and rates of change as well!

5. SKILLED WORKER'S PAY

A firm hires an unskilled worker and trains him. During training the worker's productivity is low but increases gradually. If the firm is a profit maximizer, then the cumulative pay to the worker over time must equal his cumulative value marginal product or the worker is not hired in the first place. If the firm can manage to pay the worker his value marginal product at each point of time and hires him then the above condition is satisfied. However, the facts that the worker is hired and the firm maximized profits do not imply that the firm pays the value marginal product at each point in time. Thus it is possible for the firm to pay more than value marginal product during the training and less than value marginal product afterwards and still maximize profits. Such arrangements are not exactly unheard of in labor practices.

The above assertion has been made by Gary Becker in (1980) without proof. The proof is straight forward but may be instructive and we present it here. One more excuse for presenting this formal model is to enhance the view of labor as symmetric to capital in the eyes of the firm. The institutional vehicle here is a labor contract for a "reasonable" time period.

We classify skills into m grades. Assume the firm only hires unskilled workers and that by staying with the firm they attain higher grades of skills. We assume that there are no training costs and no failures, deaths, or resignations except by retirement. Let $x_0(t)$ be the stock of unskilled workers. Then the rate of change in x_0 is given by

(1.1) $$\dot{x}_0 = u(t) - u(t - \tau_1),$$

where $u(t)$ is the unskilled workers hired at time t and where $u(t - \tau_1)$ is the number of unskilled workers hired τ_1 years ago. These workers graduate into the next grade of skill so τ_1 is the grade 1 gestation period. If we let x_i denote the stock of grade i skilled labor and τ_i denote the gestation period for that grade then the rate of change in x_i is given by:

(1.2) $$\dot{x}_i = u(t - \sum_{k=1}^{i} \tau_k) - u(t - \sum_{k=1}^{i+1} \tau_k), \qquad i = 1, ..., m.$$

Finally let x_{n+1} be the stock of retired workers and assume that they live σ years after retirement. Then the rate of change in x_n is given by:

(1.3) $$\dot{x}_{n+1} = u(t - \sum_{k=1}^{i+1}) - u(t - \sum_{k=1}^{i+1} \tau_k - \sigma).$$

In case $n = 1$, we may think of the above as a three generations model of the labor force which is suited for the study of questions of social security. On the other hand it is easy to think of each x_i as a vector of skills in case one likes to study a labor force differentiated by grade and type.

Now let there be n capital stocks; $k_1, ..., k_m$ and suppose it takes s_j units of time to build the jth type. Then the rates of change in capital are given by:

(2) $$\dot{K}_j = I_j(t - s_j),$$

where I_j is investment level in capital of type j and where we assume no depreciation. Let f(x, k) be the firm's production function and let $g^j(I_j)$ be the cost of investment in capital of type j. Let p be the output price and w be the wage rate vector $w = (w_0, w_1, ..., w_n, w_{n+1})$, where w_{n+1} is retirement income rate. Then the firm's discounted net cash flow is:

(3) $$\int_0^T [pg(x, k) - wx - \sum g^j(I_j)]e^{-rt}\, dt,$$

where r is the discount rate.

Using the results of section 2.2 above we obtain the necessary conditions for the maximization of the firm's net cash flow (3) subject to (1) and (2). The conditions are the existence of functions q_i and R_j, i = 1, ..., m; j = 1, ..., n such that:

(4.1) $$\dot{q}_i = (w_i - pf_{x_i})e^{-rt}, \qquad i = 0, 1, ..., n.$$

(4.2) $$\dot{q}_{n+1} = w_{n+1}e^{-rt}$$

(4.3) $$\dot{R}_j = -pf_{k_j}e^{-rt}, \qquad j = 1, ..., n.$$

(5) $$q_0(t) - q_0(t + \tau_1) + \sum_{i=1}^{n} [q_i(t + \sum_{k=1}^{i} \tau_k) - q_i(t + \sum_{k=1}^{i+2} \tau_k)] +$$

$$+ q_{n+1}(t + \sum_{k=1}^{n+1} \tau_k) - q_{n+1}(t + \sum_{k=1}^{n+1} \tau_k + \sigma) = 0.$$

(6) $$R_j(t + s_j) = g'^j..$$

The functions q_i and R_j are the imputed prices of labor and of capital. For that to be clearer we integrate (4) and use (5) and (6) to get:

$$(7) \quad \int_{t}^{t+\tau_1} (pf_{x_0} - w_0)e^{-r\theta}\, d\theta + \int_{t+\tau_1}^{t+\tau_1+\tau_2} (pf_{x_1} - w_1)e^{-r\theta}\, d\theta + \cdots +$$

$$\int_{t+\tau_1+\cdots+\tau_n}^{t+\tau_1+\cdots+\tau_{n+1}} (pf_{x_n} - w_n)e^{-rt} - \int_{t+\tau_1+\cdots+\tau_{n+1}}^{t+\tau_1+\cdots+\tau_{n+1}+\sigma} w_{n+1}e^{-r\theta}\, d\theta = 0.$$

$$(8) \quad \int_{t+s_j}^{T} pf_{k_j}e^{-r\theta}\, d\theta = g'^j.$$

It can easily be seen that if $w_i = pf_{x_i}$ then equation (7) is satisfied. However, the fact that equation (7) is satisfied does not imply that $w_j = pf_{x_j}$. It is indeed consistent with profit maximization to overpay earlier on and under pay later on to cover both the training and retirement period overpayment.

APPENDIX

CONSTRAINED EXTREMA OF DIFFERENTIABLE FUNCTIONALS A SURVEY

1. INTRODUCTION

In this Appendix we review the theorems that characterize optimality by way of derivatives. First we formulate a very general optimization problem. Then we present characterization theorems for three types of problems: Finite dimensional, variational, and problems in topological linear spaces. In each case we present theorems for equality - inequality constraints. The theorems in each case are: first order necessary conditions, first order sufficient conditions, second order necessary conditions, and second order sufficient conditions.

The basic technique of proofs, for the results we review, of necessity theorems is that of comparing two sets of "variations". Start at a point of a presumed optimum, and consider increments in the variables. These increments are variations. The first set of variations is that of "admissible" variations, i.e. those that satisfy the constraints in some sense. The second set of variations is that of prohibited variations, i.e. the variations that provide the maximand with a value greater than the maximum. The intersection of these sets is empty. The same can be asserted about sets of "first" and "second" variations. The conclusions of the necessity theorems consist of expressing, in terms of derivatives of maximand and constraints, the non-intersection of the two sets of variations. One way of accomplishing this is to use an implicit function theorem, e.g. Bliss, (1930), (1938), Goldstine (1940). Another way is to use a separation theorem, e.g. McShane (1939) and (1941) and Hurwicz (1958). The general approach was nicely exposited by Dubovskii and Milyutin (1965). In sufficiency theorems, local or global concavity, together with the vanishing of the first derivative of the "Lagrangian" is utilized to compare the values of the maximand at the critical point and at some other point.

The scheme of representation is as follows: Statements of theorems are followed by remarks referring the reader to the earliest, known to us, proofs of the theorems. In some instances, slight generalizations of some theorems appear here for the first time, and indications of necessary modifications to existing proofs are presented. A case is "solved" if proofs for all the four types of characterization theorems exist. The only "unsolved" case is that of problems in linear topological spaces with inequality and with equality - inequality constraints. For this case, we present two conjectures about second order conditions that are analogous to the equality constraint case. We now state the "general" optimization problem:

Let A, B, C, D be real linear topological spaces. Consider the functions:

$$f: A \to B$$
$$g: A \to C$$
$$h: A \to D$$

Let P be a partial ordering of B, let $\geq$ be defined on C - as usual - by way of convex cone and let $\theta_1, \theta_2, \theta_3$, and θ_4 be the neutral element of addition - zero elements - of A, B, C, and D respectively. The problem may be stated as follows[1]:

Find $\hat{x} \in A$ such that $f(\hat{x})$ is P-maximal[2] subject to:

$$g(x) = \theta_3$$
$$h(x) \geq \theta_4$$

In case B is the real line, P is the relation "≥" defined for real numbers and we have a problem of scalar optimization. In case B is finite dimensional we have a finite-vector maximization problem. In general, characterizations of solutions to finite-vector maximization problems may be derived from characterizations of solutions of scalar maximization problems. We shall restrict our presentation to scalar maximization problems. As an application of scalar maximization theorems we may "solve" a particular finite-vector maximization

---

[1]This general statement of the problem is due to Hurwicz (1958).

[2]We shall restrict our attention to maximization. Characterizations of solutions to minimization problems follows trivially from maximization theorems.

problem, namely for the case where P is taken to be a Pareto[1] ordering of B. For infinite-vector maximization problems, a scalarization theorem appears to be the most appropriate intermediate step to derive characterization theorems. Such a theorem was proved by Hurwicz (1958). The method, for finite-vector maximization, consists[2] of simply observing that the problem is equivalent to a finite number of scalar maximization problems. However, we shall only be concerned with scalar maximization problems.

Finally, at the risk of belaboring the obvious, three things should be pointed out. First, the choice of differentiable functions as the class to which the results apply does not mean that we think them to be the most general class of functions nor does it mean that the proofs of the theorems are the most elegant. In fact, such considerations are traded for unity and simplicity of the representation. Second, the division of problems to Finite dimensional, variational, and problems in Linear Topological Spaces. Had it been true that the problem of Part 3 of this appendix is "solved", there might be a case for disregarding sections 1 and 2, for then, they would be of only historical interest. But even then, one would be faced with the problem of assuming too much in order to get the more general results and obtain others as special cases. An illustration of the point is the relation between optimal control theory, Part 2, and Part 3, functional analysis. There exist excellent works relating the problem of maximizing a functional subject to functional constraints to optimal control problems, see e.g. Blum (1965) and Neustadt (1966). We decided not to include these papers in our survey because the results, of functional analysis, were rather specialized, i.e. they apply to functionals in general but they are more suited to variational problems. One other reason for not dealing with variational problems as problems in Linear Topological Spaces is that the theory, in Part 3, is incomplete, while variational theory is.

Third, since we are only concerned with characterization, computational results and existence theory are not reported on.

---

[1]In that case, x is said to be Pareto superior to y if $f^i(x) \geq f^i(y)$ for $i = 1, \ldots, r$ (r is the finite dimension of B), and x is Pareto optimal (P - maximal) if there does not exist a point $y \in A$ satisfying $g(y) \geq \theta_2$, $h(y) = \theta_3$ which is Pareto superior to x.

[2]Suggested to the author by Hurwicz, see Section 2, Chapter 3 of El-Hodiri (1969).

2. FINITE DIMENSIONAL PROBLEMS

In this part we let $A = E^n$, $B = E^1$, $C = E^m$, and $D = E^\ell$, where E^n, E^r, E^m, E^ℓ are Euclidean spaces of dimensions n, one, m, and ℓ respectively. For this part we reformulate the scalar maximization problem as follows:

Problem 1. Find $\hat{x} \in E^n$ such that $f(\hat{x}) \geq f(x)$ for all x satisfying $g^\alpha(x) = 0, \alpha = 1, ..., m, h^\beta(x) \geqq 0, \beta = 1, ..., \ell$.

We further state some properties of the constraints that will be used in discussing the theorems in this section. Some of the nomenclature, designated by " ", is Karush's (1939).

(D.1) Definition: Effective constraints. Let $\overset{o}{x}$ be a point that satisfies $h^\beta(\overset{o}{x}) \geq 0$, $\beta = 1, ..., m$. Let $\Gamma(\overset{o}{x})$ be the set of indices β such that $h^\beta(\overset{o}{x}) = 0$. The constraints with indices $\beta \in \Gamma(\overset{o}{x})$ will be referred to as the effective constraints at $\overset{o}{x}$.

(D.2) Definition: "*Admissible Direction*". Let h be differentiable. We say that λ is an admissible direction if λ is a non-trivial solution to the inequalities:

$$\sum_{i=1}^{n} \overset{o}{h}{}_i^\beta \lambda_i \geq 0, \qquad \beta \in \Gamma(\overset{o}{x})$$

where $\overset{o}{h}{}_i^\beta = \dfrac{\partial h^\beta}{\partial x_i} \mid x = x^0$.

(D.3) Definition. "A curve issuing from $\overset{o}{x}$ in the direction λ." By that we mean, an n-vector valued function, $\xi(t)$, of a real variable t such that $\xi(0) = \overset{o}{x}$ and $\xi'(0) = \frac{d}{dt}\xi(t)|_{t=0} = \lambda$.

(D.4) Definition. "*An admissible arc issuing from* $\overset{o}{x}$ *in the direction* λ" is an arc issuing from $\overset{o}{x}$ in the direction λ such that $h(\xi(t)) \geq 0$.

(D.5) Definition. "*Property Q" for inequality constraints*, is satisfied at $\overset{o}{x}$.

(D.6) Definition. *The rank condition for inequality constraints.* We say the rank condition for inequality constraints is satisfied at $\overset{\circ}{x}$ if and only if 1) the constraint are differentiable and 2) the rank of the matrix $[\overset{\circ}{h}{}^{\beta}_{i}]$, $\beta \in \Gamma(\overset{\circ}{x})$, equals the number of effective constraints, i.e., is maximal.

(D.7) Definition. *The rank condition for equality - inequality constraints* is satisfied at $\overset{\circ}{x}$ if and only if 1) the functions g and h are differentiable and 2) the rank of the matrix $\begin{bmatrix} \overset{\circ}{g}{}^{\alpha}_{i} \\ \overset{\circ}{h}{}^{\beta}_{i} \end{bmatrix}$, $\beta \in \Gamma(\overset{\circ}{x})$ equals $\ell +$ the number of effective constraints at $\overset{\circ}{x}$, where $\overset{\circ}{g}{}^{\alpha}_{i} = \frac{\partial g^{\alpha}}{\partial x_i}\Big|_{x = \overset{\circ}{x}}$.

(D.8) Definition. The rank condition for equality constraints is satisfied at $\overset{\circ}{x}$ if the rank of the matrix $[\overset{\circ}{g}{}^{\alpha}_{i}]$ in m.

(D.9) Definition. $\hat{x}$ is said to be a *local solution to problem 1* if and only if: There exists a neighborhood of $\hat{x}$, $N(\hat{x})$, such that $f(\hat{x}) \geq f(x)$ for all $x \in N(\hat{x})$ satisfying $g(x) = 0$ and $h(x) \geq 0$.

(D.10) Definition. $\hat{x}$ is said to be a *global solution to problem 1* if and only if $f(\hat{x}) \geq f(x)$ for all x satisfying $g(x) = 0$ and $h(x) \geq 0$.

2.1 FIRST ORDER NECESSARY CONDITIONS

Theorem 1. *If* f, g *and* h *are continuously differentiable and if* $\hat{x}$ *is a global solution to problem 1, then there exists a vector* $(\lambda_0, v, \mu) = (\lambda_0, v^1, \dots, v^m, \mu^1, \dots, \mu^{\ell}) \neq 0$ *such that,* $\lambda_0 \geq 0$ *and;*

1) $\mu^{\beta} \geq 0$, $\mu^{\beta} h^{\beta}(\hat{x}) = 0$.

2) $\hat{F}^0_x = 0$, *where* $F^0 = \lambda_0 f + v \cdot g + \mu \cdot h$, *and* F_x *is the vector of partial derivatives of* F *with respect to the components of* x *evaluated at* $x = \hat{x}$.

For the case of equality constraints, i.e. the set $\{x \mid h(x) \geq 0\} = E^n$, Theorem 1 was proved by Caratheodory[1] (1935), Bliss (1938) and McShane (1941).[2] For the case of inequality constraints, i.e. the set $\{x \mid g(x) = 0 = E^n$, Theorem 1 - except for the non-negativity of μ in condition 1 of the conclusion - was proved by Karush[3] (1939). The non-negativity of μ may be proved by using a separation theorem[4]. This was, essentially, the crux of Fritz John's (1948) proof who has treated a more general problem of an infinite number of inequality constraints.

Theorem 2. *If, in addition to the assumptions of theorem 1 we have either:*[5]
a) *Property Q for inequality constraints,* D.5, *and the rank condition for equality constraints,* D.7, *are satisfied at* $\hat{x}$.
b) *The rank condition for equality-inequality constraints is satisfied at* $\hat{x}$.
Then the conclusions of theorem 1 follow with $\lambda_0 \neq 0$.

For equality constraints the theorem was proved as a corollary of theorem 1 by Bliss[6] (1938). For the case of inequality constraints the theorem was proved by Karush[7] (1939). The proof, for equality-inequality constraints, may be accomplished by converting inequality constraints to equality constraints and obtaining the theorem as a

---

[1]Theorem 2, section 187 part II.
[2]Theorem 1.1 of Bliss (1938). In McShane (1941) utilizes a separation theorem to characterize the problem with equality constraints.
[3]Theorem 3.1.
[4]Separating the linear sets: $\hat{f}_i \xi_i < 0$ and $\overset{\circ}{h}{}^{\beta}_i\, \xi_i \geq 0$. See Dubovskii and Milyutin (1965) for an extensive discussion.
[5]The conditions that follow are alternative forms of the constraint qualification, see Arrow-Hurwicz, Uzawa (1955) for other forms and for relations among various forms of the constraint qualification.
[6]In remarks following theorem 1.1.
[7]Theorem 3.2.

corollary to theorem 1. This was presented by Pennisi[1] (1953). A direct proof was presented by Hestenes[2] (1965).

2.2. FIRST ORDER SUFFICIENT CONDITIONS

Theorem 3. *If:* 1) f, g, *and* h *are differentiable* 2) *The conclusions of theorem 1 hold with* $\lambda_0 > U$ *at a point* $\hat{x}$ *with* $g(\hat{x}) = 0$, $h(\hat{x}) \geq 0$. 3) Either (3.a) F^0, *of theorem 1, is concave or* (3.b) g *is linear and* h *is concave. Then* $\hat{x}$ *is a global solution to problem 1.*

The theorem follows from the fact that a concave function lies below its tangent plane. The implications of this fact were utilized by Kuhn-Tucker (1953) in the proof of their equivalence theorem. The present theorem may be proved by applying Lemma 3 of Kuhn and Tucker (1953) to F^0.

2.3. SECOND ORDER NECESSARY CONDITIONS

Theorem 4. *If:* 1) f, g, *and* h *have second order continuous partial derivatives.* 2) $\hat{x}$ *is a solution to problem 1.* 3) *The rank condition for equality-inequality constraints,* D.6, *is satisfied. Then there exist multipliers* $(v, \mu) = (v^1, ..., v^m; \mu^1, ..., \mu^l)$ *such that: The conclusions of theorem 2 hold and* $\Sigma_{i,j}\hat{F}_{ij}\eta_i\eta_j \leq 0$, *for all* $\eta = (\eta_1, ..., \eta_n) \neq 0$ *with:* $\Sigma_i\hat{g}_i^\alpha\eta_i = 0$, $\Sigma_i\hat{h}_i^\beta\eta_i = 0$, $\alpha = 1, ..., m$, $\beta \in \Gamma(\hat{x})$ *where,* $F = f + \sum_\alpha v^\alpha g^\alpha + \sum_\beta \mu^\beta h^\beta$ *and* $\hat{F}_{ij} = \frac{\partial^2 F}{\partial x_i \partial x_j} |x = \hat{x}$.

For the equality constraints, the theorem was proved by Caratheodory[3] (1935) and Bliss[4] (1938). For inequality constraints the

---

[1]Theorem 3.1.
[2]Theorem 10.1, Chapter 1.
[3]Theorem 3, section 212, Part II.

theorem was proved by Karush[1] (1939). For the equality-inequality case, the theorem was proved by Pennisi[2] (1953), under the, dispensable[3], stipulation that the number of zero multipliers attached to effective inequality constraints is at most one. A direct proof was presented by Arrow and Hurwicz, (1955), theorem 3.

2.4. SECOND ORDER SUFFICIENT CONDITIONS

Theorem 5. *If:* 1) f, g, *and* h *have continuous second order derivatives* 2) *The conclusions of theorem* 2 *holds at* $\hat{x}$ *with* $g(\hat{x}) = 0$, $h(\hat{x}) > 0$. 3) *The conclusions of theorem 4 holds with strict inequality for* $\eta \neq 0$. *Then* $\hat{x}$ *is a local solution to problem 1.*

For equality constraints the theorem was proved by Bliss[4] 91938). Caratheodory[5] (1935) assumes, needlessly, that the rank condition holds. A closely related theorem was proved by Pennisi[6] (1953), where the restrictions on η are augmented by requiring that $\Sigma_i \hat{h}_i^{\beta} \eta_i \geq 0$ for indices β with $\mu^{\beta} > 0$. Theorem 5 may be proved by applying the sufficiency theorem for equality constraints by using Karush's device[7] (1939). A direct proof was provided by Hestenes[8] (1966).

---

[4]Theorem 1.2.
[1]Theorem 5.1.
[2]Corollary to theorem 3.2.
[3]See El-Hodiri (1969), where Pennisi's theorem is proved by directly applying second order necessary conditions for equality constraints using Karush's device.
[4]Theorem 1.3.
[5]Theorem 4, section 213, Part II.
[6]Theorem 6.1.
[7]See 12).
[8]Theorem 10.3, chapter 1.

2.5. SECOND ORDER CONDITIONS IN TERMS OF DETERMINANTS

Let the matrix $H = \begin{bmatrix} H_1 & H_2 \\ H_3 & H_4 \end{bmatrix}$ be defined as follows: $H_1 = [\hat{F}_{ij}]$, where $\hat{F}_{ij}$ is as defined in theorem 4, $H_2 = H_3' = \begin{bmatrix} \hat{g}^{\alpha}_{i} \\ \hat{h}^{\beta}_{i} \end{bmatrix}$, $\beta \in \Gamma(\hat{x})^2$, with primes denoting transposes and using the notation of definition D.7 and where H_4 is a square matrix of zeros of order $b = \ell$ + the number of effective constraints at \o9x,^). Let $h^{(k)}$ denote *bordered principle minors of order* k *of* H defined as follows: 1) Eliminate all rows and columns of H_1 except the first k, 2) Eliminate all rows of H_2 except the first k_1, 3) Eliminate all columns of H_3 except for the first k, 4) Keep H_4 as it is.

Corollary to Theorem 4: *Under the assumption of theorem 4, the bordered principle minors of* H *of orders* k, *ranging from* $b = \ell$ + *the number of effective constraints at* $\hat{x}$ *to* n, *alternate in signs. Some or all may be zero. The first bordered principle minor, of order* b, *has the sign of* $(-1)^{b+1}$.

Corollary to Theorem 5: *Condition 3 of theorem 5 may be replaced by: None of the bordered principle minors of* H *of order* $b < r < n$ *is zero and they alternate in sign with the first, of order* b, *having the sign of* $(-1)^{b+1}$.

Both corollaries follow from Caratheodory's results, sections 209, 213-217 of (1935), on necessary and sufficient conditions for a quadratic form to be semi-definite (definite) under linear constraints.

3. A VARIATIONAL PROBLEM

In view of the vast literature on the calculus of variations and wide accessibility of such literature, this part will be very brief. Many interesting topics will be left out, e.g. problems with retarded arguments[1]. For a survey of variational problems with equality

[1] See El'sgol'c (1955), Halanay (1968), and Ewing (1969).

constraints, the reader is referred to Bliss (1930), which we shall take as a point of departure for this part. We formulate the Bolza-Hestenes[1] problem in the Calculus of variations as follows: Let T be a subset of the non-negative half of the real line. Consider the class of piecewise smooth functions $x(t)$ defined on t and having values in E^n together with their derivatives $\dot{x}$. The problem is:

Problem 2. Find $\hat{t}_0, \hat{t}_1, \hat{x}, \hat{x}(\hat{t}), \hat{x}(\hat{t}_0), \hat{x}(\hat{t}_1) = \hat{Z}$ that maximizes $J^0[Z] = \int_{t_0}^{t_1} f^0(t, x, \dot{x})\, dt + g^0(t_0, t_1, x(t_1))$ subject to

(1) $J^{\alpha}[Z] = \int_{t_0}^{t_1} f^{\alpha}(t, x, \dot{x})\, dt + g^{\alpha}(t_0, t_1, x(t_0, x(t_1)) = 0, \qquad \alpha = 1, ..., m_1,$

(2) $J^{\bar{\alpha}}[Z] = \int_{t_0}^{t_1} f^{\bar{\alpha}}(t, x, \dot{x})\, dt + g^{\bar{\alpha}}(t_0, t_1, x(t_0, x(t_1)) = 0, \qquad \bar{\alpha} = m_1 + 1, ..., m,$

(3) $\phi^{\beta}(t, x, \dot{x}) = 0, \qquad \beta = 1, ..., \ell_1,$

(4) $\phi^{\bar{\beta}}(t, x, \dot{x}) \geq 0, \qquad \bar{\beta} = \ell_1 + 1, ..., \ell,$

(5) $\psi^{\gamma}(t_0, t_1, x(t_0), x(t_1)) = 0, \qquad \gamma = 1, ..., S_1,$

(6) $\psi^{\bar{\alpha}}(t_0, t_1, x(t_0), x(t_1)) \geq 0, \qquad \bar{\gamma} = S_1 + 1, ..., S,$ where $t_0, t_1 \in T$ with $t_0 < t_1$.

As was noted, by Berkovitz (1961) and Hestenes (1965), the above problem is equivalent to the problem of optimal control[2]. The results of optimal control theory are derivable from the results that we shall present.[3]

---

[1] See Hestenes (1965).

[2] As stated in these two papers.

[3] See Berkovitz (1961) and Hestenes (1965) for the necessary transformations.

3.1. FIRST ORDER NECESSARY CONDITIONS

Theorem 6. *If:* 1) *The functions* f, g, ϕ, *and* ψ *are continuously differentiable, as functions of real variables.* 2) $\hat{Z}$ *is a solution to problem 2.* 3) *The matrix* $\begin{bmatrix} \hat{\phi}^{\beta}_{x_i} \\ \hat{\phi}^{\bar{\bar{\beta}}}_{x_i} \end{bmatrix}$ *has full rank at each point* $t \in [\hat{t}_0, \hat{t}_1]$, *where* $\hat{\phi}^{\beta}_{x_i} = \left.\frac{\partial \phi^{\beta}}{\partial \dot{x}_i}\right|_{Z=\hat{Z}}$ $\hat{\phi}^{\bar{\bar{\beta}}}_{x_i} = \left.\frac{\hat{\phi}^{\bar{\beta}}}{\partial \dot{x}_i}\right|_{Z=\hat{Z}}$ *and where the index* $\bar{\beta}$ designates *indices* β *with* $\phi^{\bar{\beta}}(\hat{Z}) = 0$, i = 1, ..., n. *Then there exist a constant vector* $(\lambda_0, q, \omega) = \Lambda_0; q^1, ..., q^m; \omega^1, ..., \omega^s)$ *and a vector function* $p = (p^1, ..., p^{\ell})$ *defined on* $[\hat{t}_0, \hat{t}_1]$ *such that*

1) *There does not exists* $t \in [\hat{t}_0, \hat{t}_1]$ *with* $[\lambda_0, q, \omega, p] = 0$, *also* $(\lambda_0, q, \omega) \neq 0$

2) $\Lambda_0 \geq 0$, $q^{\bar{\alpha}} \geq 0$, $q^{\bar{\alpha}} J^{\bar{\alpha}}[Z] = 0$

3) P(t) *is piecewise continuous, continuous at points of continuity of* $\dot{x}$, $P^{\bar{\beta}}(t) \geq 0$, $P^{\bar{\beta}}(t)\phi^{\alpha}[\hat{Z}] = 0$

4) $\omega^{\gamma} \geq 0$, $\omega^{\gamma}\psi^{\gamma}[\hat{Z}] = 0$

5) $\frac{d}{dt}\hat{F}_{\dot{x}_i}$, i = 1, ..., n, *where* $F = \lambda_0 f^0 + \sum_{\alpha} q^{\alpha} f^{\alpha} + \sum_{\beta} \mu^{\beta}\phi^{\beta}$, $F_{\dot{x}_i} = \left.\frac{\partial F}{\partial \dot{x}_i}\right|_{Z=\hat{Z}}$

and $\hat{F}_{x_i} = \left.\frac{\partial F}{\partial \dot{x}_i}\right|_{Z=\hat{Z}}$

6) $\frac{d}{dt}(\hat{F} - \sum_i \hat{\dot{x}}_i \hat{F}_{\dot{x}_i}) = \frac{\partial}{\partial t}\hat{F}.$

7) (Transversality conditions)

$d\,G + [(\hat{F} - \sum_i \hat{\dot{x}}_i F_{\dot{x}_i})\, dt_{\gamma} + \sum_i \hat{F}_{\dot{x}_i}\, dx_i(t_{\gamma})]_{\gamma=0}^{\gamma=1} = 0$ *is an identity in* $dx_i(t_{\gamma})$, dt_{γ}, $\gamma = 0, 1$; i = 1, ..., n; *where* $G = \lambda_0 g^0 + \sum_{\alpha} q^{\alpha} g^{\alpha} + \sum_{\gamma} \omega^{\gamma}\psi^{\gamma}$,

8) (Weierstrass)

$$E(t, \hat{x}, \hat{\dot{x}}, \dot{x}) = F(t, \hat{x}, \hat{\dot{x}}) - F(, \hat{x}, \dot{x}) - \sum_i (\hat{\dot{x}}_i - \dot{x}_i)\hat{F}_{\dot{x}_i} \geq \sum_{\overline{\beta}} \mu^{\overline{\beta}} \phi^{\overline{\beta}}.$$

whenever $(t, \hat{x}, \dot{x})$ *satisfy the constraints (1) - (6).*

For equality constraints 1) - 7) were proved by Bliss (1930) and 8) was proved by McShane (1939) and (1941) with a new proof for 1) - 8) utilizing separation of convex sets. See Bliss (1930) for further references.

For fixed end points (i.e. t_0, t_1, $x(t_0)$ and $x(t_1)$ are constants), Valentine (1937) proved that conclusion: 5) and 8) are necessary with conclusion 3) holding for p except for the non-negativity of $P^{\overline{\beta}}$. The non-negativity of $P^{\overline{\beta}}$ was proved by applying a Clebsch[1] type second order necessary condition[2]. Valentine's method consisted of converting differential inequality constraints to equations by subtracting from each, the square of a derivative of an added variable. Then he derived his results as applications of characterization theorems for the problem of Bolza with equality constraints[3]. For the general problem we may convert the inequality constraints (2) and (6) to equality constraints by introducing three new sets of variables $y_1^{\overline{\alpha}}(t)$, $y_2^{\overline{\beta}}(t)$, and $y_3^{\overline{\tau}}(t)$ as follows: $\dot{y}_1^{\overline{\alpha}}(t) = 0$, $y_1^{\overline{\alpha}}(t)$ free, $\dot{y}_2^{\overline{\tau}}(t) = 0$, $y_2^{\overline{\beta}}(t)$ free and $\dot{y}_3^{\overline{\tau}}(t) = 0$, $y_3^{\overline{\tau}}(t)$ free, and consider equivalent problem of maximizing J^0 subject to constraints (1), (3), (5) and: (2^1) $\overline{J}^{\overline{\alpha}} = J^{\overline{\alpha}} - (\overline{y}_1^{\overline{\alpha}}(t_1))^2 = 0$, $(4)^1$ $\overline{\phi}^{\overline{\beta}} = \phi^{\overline{\beta}} - (\dot{y}_2^{\overline{\beta}})^2 = 0$ and (6') $\overline{\psi}^{\overline{\tau}} = \psi^{\overline{\alpha}} - (y_3^{\overline{\tau}}(t_1))^2 = 0$. We then get, in addition to Valentine's conditions, condition 7) of theorem 6. Noting that condition 6) may be obtained from 5) and 7) we would have all of the conditions of the theorem. A direct elegant proof of theorem 6 was provided by Hestenes (1965).

---

[1]See section 3.3 in this appendix.

[2]Corollary 3.4.

[3]These theorems may be found in Bliss (1930).

We now discuss a condition that guarantees that λ_0 in theorem 6 is non-zero. The case where the multipliers are unique, choosing $\lambda_0 = 1$, is what is known in the literature as the normal case. Although conditions for normality are very hard to verify in applications, we shall present one of these conditions in this section[1].

(D.11) Definition. *Normality.* $\hat{Z}$ is said to be normal if and only if conclusions 1) - 5) and 7) of theorem 6 hold with $\lambda_0 = 1$ and the multipliers q, p, ω are unique.

(D.12) Definition. *Admissible variations.* Consider a point Z and a vector valued functions $\xi^\sigma(t) = (\xi_1^\sigma(t), ..., \xi_n^\sigma(t))$; $\sigma = 1, ..., \bar{m} + \bar{s}$, where $\bar{m} = m_1$ + the number of constraints J^α that hold as equations at Z, effective at the end points of Z, and where $\bar{S} = S$ + the number of constraints ψ^τ that are effective at Z. $\xi(t) = (\xi^1(t), ..., \xi^{\bar{m}+\bar{s}}(t))$ is said to be admissible variations if and only if:

1) $\xi_i(t)$ are differentiable on $[\bar{t}_0, \bar{t}_1]$,

2.1) $\Phi^\beta(\xi^\sigma) = \sum_{i=1}^{m} \bar{\phi}^\beta_{x_i} \xi_i^\sigma(t) + \sum_{i=1}^{m} \bar{\phi}^\beta_{\dot{x}_i} \dot{\xi}_i^\sigma(t) = 0$, $\beta = 1, ..., m_1$,

2.2) $\Phi^{\bar{\bar{\beta}}}(\xi^\sigma) = \sum_i \bar{\phi}^{\bar{\beta}}_{x_i} \xi_i^\sigma + \sum_i \bar{\phi}^{\bar{\beta}}_{\dot{x}_i} \dot{\xi}_i^\sigma(t) = 0$, for $\bar{\beta}$ with $\phi^{\bar{\bar{\beta}}}(Z) = 0$, where

$$\bar{\phi}^\beta_{x_i} = \left.\frac{\partial \phi^\beta}{\partial x_i}\right|_{Z = \bar{Z}}$$

and where $\bar{\phi}_{\dot{x}_i}$, $\bar{\phi}^{\bar{\bar{\beta}}}_{x_i}$, and $\bar{\phi}^{\bar{\bar{\beta}}}_{\dot{x}_i}$ are defined similarly.

(D.13) Definition. The rank condition. The first rank condition is said to be satisfied at Z if there exists a set of admissible variations

[1] See Berkovitz (1961), section VIII, theorem 3, for alternative sufficient condition for normality.

ξ^σ, and arbitrary constants $\tau_0^\sigma, \tau_1^\sigma$ $\sigma = 1, ..., \bar{m} + \bar{s}$, such that the matrix $\begin{bmatrix} L_1 \\ L_2 \\ L_3 \\ L_4 \end{bmatrix}$ has rank $\bar{m} + \bar{s}$, where

$$L_1 = [L_1^{\alpha\sigma}] = [(g^\alpha_{t_0} + \sum_i g^\alpha_{\dot{x}_{i(t_0)}} \dot{\bar{x}}_1(t_0) - f^\alpha(\bar{t}_0, \bar{x}(\bar{t}_0), \dot{\bar{x}}(t_o)))\, \tau_0^\sigma + \sum_i \bar{g}^\alpha_{x_{i(t_0)}} + \xi_i^\sigma(t_0) + (\bar{g}^\alpha_{t_1} + \sum_i \bar{g}^\alpha_{x_i}(t_1) + f^\alpha(\bar{t}_1, \bar{x}(\bar{t}_1), \dot{\bar{x}}(\bar{t}_1)))\, \tau_1^\sigma + \sum_i \bar{g}^\alpha_{x_{i(t_1)}} + \xi_i^\sigma(t_1)$$

$$+ \int_{\bar{t}_0}^{\bar{t}_1} (\sum_i f^\alpha_{x_i}\xi_i^\sigma(t) + \bar{f}^\alpha_{\dot{x}_i}\dot{\xi}_i^\sigma(t))\, dt],\ \alpha = 1, ..., m_1,\ \sigma = 1, ..., \bar{m} + \bar{s},$$

$$L_2 = [L_2^{\bar{\alpha}\sigma}] = [\bar{g}^{\bar{\alpha}}_{t_0} + \sum_i \bar{g}^{\bar{\alpha}}_{\dot{x}_{i(t_0)}} \dot{\bar{x}}_1(t_0) - f^{\bar{\alpha}}(\bar{t}_0, \bar{x}(\bar{t}_0), \dot{\bar{x}}(t_o)))\, \tau_0^\sigma + \sum_i \bar{g}^{\bar{\alpha}}_{x_{i(t_0)}} + \xi_i^\sigma(t_0) + (\bar{g}^{\bar{\alpha}}_{t_1} + \sum_i \bar{g}^{\bar{\alpha}}_{x_i}(t_1) + f^{\bar{\alpha}}(\bar{t}_1, \bar{x}(\bar{t}_1), \dot{\bar{x}}(\bar{t}_1)))\, \tau_1^\sigma + \sum_i \bar{g}^{\bar{\alpha}}_{x_{i(t_1)}} + \xi_i^\sigma(t_1)$$

$$+ \int_{\bar{t}_0}^{\bar{t}_1} (\sum_i \bar{f}^\alpha_{x_i}\xi_i^\sigma(t) + \bar{f}^\alpha_{\dot{x}_i}\dot{\xi}_i^\sigma(t))\, dt],$$ $\bar{\bar{\alpha}}$ denotes indices of effective constraints $J^\sigma[Z]$ at $Z, \bar{\sigma} = 1, ..., \bar{m} + \bar{s}$,

$$L_3 = [L_3^{\gamma\sigma}] = [\bar{\psi}^\alpha_{t_0} + \sum_i \bar{\psi}^\gamma_{\dot{x}_{i(t_0)}} \dot{\bar{x}}_1(t_0)\, \tau_0^\sigma + \sum_i \bar{\psi}^\gamma_{x_{i(t_0)}} + \xi_i^\sigma(t_0) + (\bar{\psi}^\gamma_{t_1} + \sum_i \bar{\psi}^\gamma_{x_i}(t_1)\, \dot{\bar{x}}(\bar{t}_1)))\, \tau_1^\sigma + \sum_i \bar{\psi}^\gamma_{x_{i(t_1)}} + \xi_i^\sigma(t_1)],\ \gamma = 1, ..., s_1,\ \sigma = 1, ..., \bar{m} + \bar{s}$$

$$L_4 = [L_4^{\bar{\gamma}\sigma}] = [\bar{\psi}^{\bar{\gamma}}_{t_0} + \sum_i \bar{\psi}^{\bar{\gamma}}_{\dot{x}_{i(t_0)}} \dot{\bar{x}}_1(t_0)\, \tau_0^\sigma + \sum_i \bar{\psi}^{\bar{\gamma}}_{x_{i(t_0)}} + \xi_i^\sigma(t_0) + (\bar{\psi}^{\bar{\gamma}}_{t_1} + \sum_i \bar{\psi}^{\bar{\gamma}}_{x_i}(t_1)\, \dot{\bar{x}}(\bar{t}_1)))\, \tau_1^\sigma + \sum_i \bar{\psi}^{\bar{\gamma}}_{x_{i(t_1)}} + \xi_i^\sigma(t_1)],,$$ $\bar{\bar{\gamma}}$ denotes indices of $\psi^{\bar{\gamma}}$ that are effective at $Z, \bar{\sigma} = 1, ..., \bar{m} + \bar{s}$, and where subscripts

denote partial derivatives with respect to indicated variables and "-" above an expression indicates that it is evaluated at Z.

Theorem 7. *If* f, g, ϕ, *and* ψ *are differentiable then the rank condition, definition D.12, at* Z *is necessary and sufficient for the normality of* Z.

In the absence of inequality constraints, the theorem was proved by Bliss[1] (1930). Reformulating the problem with added inequality constraints, as indicated above, the theorem is obtained as a straight forward application of Bliss's theorem.

3.2 FIRST ORDER SUFFICIENT CONDITIONS

Theorem 8. *If* 1) f, g, ϕ, *and* ψ *are differentiable, and concave.* 2) *Conclusions 1) - 5) and 7) of theorem 6 are satisfied with:* 2: i) $\lambda_0 > 0$, 2. ii) $q^\gamma > 0$, $\omega^\gamma > 0$, $\gamma = 1, \ldots, m_1$, $\gamma = 1, \ldots, s_1$, *at a point* $\hat{Z}$ *that satisfies the constraints (1) - (6). Then* $\hat{Z}$ *is a global solution to problem 2.*

The theorem was proved by Mangasarian (1966) for the canonical, optimal control, problem with fixed t_0 and t_1. Theorem 8 may be proved by repeating the steps of Mangasarian's proof for problem 2. As Mangasarian notes, in the absence of equality constraints condition 2. ii) of theorem 8 is not needed, nor is it required if the equality constraints are linear.

3.3 SECOND ORDER NECESSARY CONDITIONS

Theorem 9. (Jacobi-Myer-Bliss) *If:* 1) f, g, h, *and* ψ *have continuous second order derivatives,* 2) $\hat{Z}$ *is a solution to problem 2 and* 3) $\hat{Z}$ *is normal. Then there exist multipliers as in theorem 6 with* $\lambda_0 = 1$ *such that:* Q(\o(Z,^), τ, ξ) = d^2(G(\o(Z,^); τ, ξ) + [(\o(F,^)$_t$ - \i\su(i,,

---

[1]Section 6, Page 693.

$$)\hat{F}_{\dot{x}_i\dot{x}_i})\tau_\gamma^2 + 2\sum_i \hat{F}_{x_i}\xi_i(\hat{t}\gamma)\,\tau_\gamma)]|_{\gamma=0}^{\gamma=1} + \int_{\hat{t}_0}^{\hat{t}_1} (\sum_{i,j} \hat{F}_{x_ix_j}\,\xi_i\,\xi_j + 2\sum_{i,j} \hat{F}_{x_i\dot{x}_j}\xi_i\dot{\xi}_j +$$

$$\sum_{i,j} \hat{F}_{\dot{x}_i\dot{x}_j}\dot{\xi}_i\dot{\xi}_j)\,dt \leq 0,$$ *for* $(\tau, \xi) = (\tau_0, \tau_1, \xi(t), ..., \xi(t)) \neq 0$ *satisfying*

(i) $\hat{\Phi}^\beta(\xi) = 0, \quad \hat{\Phi}^{\bar{\beta}}(\xi) = 0$

(ii) $\hat{L}_1^\gamma(\tau, \xi) = 0, \hat{L}_2^{\bar{\gamma}}(\tau, \xi) = 0, \hat{L}_3^\gamma(\tau, \xi) = 0, \hat{L}_4^{\bar{\gamma}}(\tau,\xi) = 0,$

with terms in (i) and (ii) defined as in (2) of D.12 and as in D.13 with (τ, ξ) *as a matrix with one row* $(\tau^\sigma, \xi^\sigma) = (\tau, \xi)$, *and where:* $d^2G(\hat{Z}; \tau, \xi)$ *is the second differential of* G *at* Z *with* (τ, ξ) *as increments, a single subscript denotes a first derivative and a double subscript denotes second (mixed partial) derivatives with "^" signifying evaluation at* $\hat{Z}$.

For equality constraints the theorem was proved by Bliss[1] (1930) and (1946)[2]. With added inequality constraint the theorem may be proved by applying Bliss's theorem to our problem, after converting it to a problem with equality constraint as we indicated in the discussion of theorem 6.

Theorem 10. (The Clebsch condition). *If the hypotheses of theorem 9 are satisfied then* $\sum_{i,j} \hat{F}_{\dot{x}_i\dot{x}_j}\Pi_i\Pi_j \leq 0$ *for* $\Pi = (\Pi_1, ..., \Pi_n) \neq 0$ *satisfying* $\sum_{i=1}^{n} \hat{\phi}^\beta_{\dot{x}_i}\Pi_i = 0$, $\beta = 1, ..., \ell_1$, $\sum_i \hat{\phi}^{\bar{\beta}}_{\dot{x}_i}\Pi_i = 0$, $\bar{\beta}$ *are indices of inequality constraints that are effective at* Z, $\sum_i \hat{f}^\alpha_{\dot{x}_i}\Pi_i = 0$, $\alpha = 1, ..., m_1$, $\sum_i \hat{f}^{\alpha}_{\dot{x}_i}\Pi_i = 0$, α *indicates constraints* J^α *effective at* $\hat{Z}$.

---

[1]Sections (24) - (26).

[2]Theorem 80, p. 228, the statement and proof here are more complete than they are in Bliss (1930).

Theorem 10 was proved by Valentine[1] (1937) for fixed end points. Valentine's method of proof can, easily, be applied to prove our theorem. See also McShane (1939) for the problem with equality constraints.

3.4 SECOND ORDER SUFFICIENT CONDITIONS

For the purposes of this section, we have to define a weak local solution of problem 2.

(D.14) Definition. *Weak Local Solution.* We say that $\hat{Z}$ is a weak local solution if $J^0[\hat{Z}] \geq J^0[Z]$ for all Z satisfying the constraints (1) - (6) with $\|(\hat{x} - x, \dot{\hat{x}} - \dot{x})\| < \varepsilon$ for some $\varepsilon > 0$, where $\|\cdot\|$ denotes the euclidian norm.

Theorem 11. (Pennisi) *If* 1) f, g, ϕ, *and* ψ *have continuous second order partial derivatives.* 2) *Conclusions 1 - 5 and 7 of theorem 6 are satisfied at a point* Z *that satisfies the constraints 1 - 6.* 3) *The matrix* $\begin{bmatrix} \hat{\phi}^{\beta}_{x_i} \\ \hat{\phi}^{\bar{\bar{\beta}}}_{x_i} \end{bmatrix}$, *of theorem 6, has full rank.* 4) *The form* $Q(Z; \tau, \xi)$ *of theorem 9 is negative definite under constraints (i) and (ii) (in the statement of theorem 9) for* $(\tau, \xi) \neq 0$. *Then there exists an* $\varepsilon > 0$ *such that* $\hat{Z}$ *is a weak solution of problem 2, in the sense of definition D.14.*

Assuming that $\lambda_0 = 1$ and that at most one inequality constraint is effective at $\hat{Z}$, Valentine[2] proved an analogous theorem in (1937). Pennisi[3] (1953) proved a sufficiency theorem without the assumptions of Valentine. Pennisi's theorem is stronger than theorem 11 in the sense that Q is negative on the subset of variations (τ, η) of theorem 11 which in addition (to (i) and (ii) of theorem 9) satisfy some

[1]Corollary 3:4 p. 9.
[2]Theorem 10.2, section 10.
[3]Theorem 2.1.

inequalities for those constraints that are effective but have zero multipliers.

One way to prove theorem 11 is to convert the problem into one with equality constraints and apply Pennisi's theorem. Another way is to note that theorem 11 is corollary of Pennisi's theorem, after modifying the latter to take care of the additional constraints. Conditions for the non-negativity of the quadratic forms in sections 3.3 and in this section, analogous to those of section 2.5, remain to be worked out.

4. A PROBLEM IN TOPOLOGICAL LINEAR SPACES

Let A, C, D be real Banach spaces and let R denote the real line. Consider f: A → R, A → C and h: A → D. The problem we study here is:

Problem 3. Maximize f(x) subject to g(x) = θ_3 and h(x) ≥ θ_4 where the inequality θ_3 and θ_4 are as defined in the introduction.

Dealing with differentiable functions we note that there are numerous equivalent[1] ways of defining derivatives in linear spaces. We shall use Frechét's definition and mean "Frechét differentiable" when we say that a function is differentiable.[2]

4.1 FIRST ORDER NECESSARY CONDITIONS

Theorem 12. *If* f, g, *and* h *are differentiable and if* $\hat{x}$ *is a solution to problem 3 and there exists a constant* $\lambda_0 \geq 0$ *and linear functionals* ℓ^1: C → R, ℓ^2: D → R *such that* 1) $\ell^2 \geq 0$, $[\ell^2, h(\hat{x})] = 0$, *where* $[\ell^2, h]$ *denotes the value of the functional* ℓ^2 *at* $h(\hat{x})$. 2) *For any* $y^1 \in C$, $y^2 \in D$, *the triple* $(\lambda_0, [\ell^1, y^1], [\ell^2, y^2] \neq 0$, 3) $F'(\hat{x}) = 0$, *where* $F = \lambda_0 f + [\ell^1, g] + [\ell^1, h]$ *and* $F'(\hat{x}) = dF(\hat{x}, \xi)$ *with* ξ *as the "increment" in the definition of the differential.*

---

[1]See Averbukh and Smolyanov (1968).

[2]See Vainberg (1956) and Liusternik and Sobolev (1951) for an exposition of calculus in linear spaces.

The theorem follows directly from theorem 2.1 in Duboviskii and Milyutin[1] (1965).

Conditions that guarantee that $\lambda_0 > 0$ and that the functionals ℓ^1 and ℓ^2 are unique are referred to in the literature, pertaining to problem 3, as regularity conditions and as constraint qualifications. We shall now list these conditions and present some sufficient conditions for them to hold.

4.1.1 REGULARITY CONDITIONS

(R.1) (Gapushkin (1964)): For equality constraints, $\bar{x}$ is said to (R.1)[2] regular if for every $\xi \in A$ with $g'(\bar{x}, \xi) = \theta_3, \xi \neq \theta_1$ we have: There exists a function of a real variable t, V: [0, 1] and $V'(0, \tau) = \xi$.

(R.2) (Hurwicz (1958)): For the inequality constraint, $\bar{x}$ is (R.2) regular if and only if: For any $\xi \in A$ with $\xi \neq \theta_1$ such that $x = \bar{x} + \xi$ implies $h'(\bar{x}, \xi) + h(x) \geq \theta_4$, we have: There exists a function of a real variable t, $V: [0, 1] \to A$ such that:

(i) $V'(t, \tau)$ exists for $t \in [0, 1]$

(ii) $\bar{x} = V(0)$

(iii) $h(V(t)) \geq \theta_4, t \in [0, 1]$

(iv) $V'(0, \tau) = \xi, \tau > 0$.

(R.3) (Gapushkin (1967)): $\bar{x}$ is said to be (R.3) regular if and only if:

For any $\xi \in A$ with $g'(\bar{x}, \xi) = \theta_3$ and $h(\bar{x}) + h'(\bar{x}, \xi) \geq \theta_4$ we have: There exists a functions of a real variable t; $V: [0, 1] \to A$ such that

(a) $v(0) = \bar{x}$

---

[1]Dubovskii and Milyutin (1965) utilize the fact that the set of "variations" that give the maximand value greater than the maximum could not intersect with the sets of "variations" that satisfy the constraints. By variations they mean differentials at x. Since these sets are defined by linear inequalities and equations they are convex. Using a separation theorem they derive what they call the Euler equation. Writing the Euler equation in terms of differentials of the maximand and constraints we obtain conclusion 3 of theorem 12.

[2]For finite dimensional spaces, this is equivalent to the rank condition.

(b) $g(v(t)) = \theta_3$, $h(V(t)) > \theta_4, t \in [0, 1]$

(c) $V'(t, \tau)$ exists for $t \in [0, 1]$

(d) $V'(0, \tau) = \xi, \tau > 0$.

Remark 1: (R.3) is a specialization of Gapushkin's regularity condition which is a uniform condition[1]. (R.1) is a further specialization for the case of equality constraints.

Remark 2: Recall that the inequality in constraint 2) of problem 3 is defined in terms of a closed convex cone, say, K_2. Let $K = \{\theta_3\} \otimes K_2$, where $\{\theta_3\}$ is a cone that contains only θ_3. Let $\theta = \theta_3 \otimes \theta_4$ and let $G: A \to C \otimes D$ be the "pair" valued function $<g, h>$. Then we may write constraints 1) and 2) in the form $G(x) \geq \theta$ where "$\geq$" is in the sense of K. With that formulation, (R.2) becomes a regularity condition for equality-inequality constraints.

4.1.2 SUFFICIENT CONDITIONS FOR REGULARITY

We now present some conditions that imply regularity. We present some sufficiency lemmas for equality constraints and some sufficiency lemmas for equality-inequality constraints. These last lemmas are, of course, sufficiency lemmas, for R.1 and R.2 regularity. However, they may be strengthened by specializing the conditions when we are concerned with (R.1) regularity or (R.2) regularity. Before stating these conditions we introduce some notations.[2]

0.1) The constraint set $N = \{x \in A: g(x) = \theta_3 \text{ and } h(x) \geq \theta_4\}$.

0.2) Let $\|\cdot\|_A$ the norm of the space A, a sphere in A with center at $\bar{x}$ and radius δ will be denoted by $\gamma(\bar{x}, \delta)$ and $\gamma(\bar{x}, \delta) = \{x: \|x - \bar{x}\|_A \leq \delta\}$.

0.3) The set $D_{\bar{x}} = \{\zeta \in A: g'(\bar{x}, \xi) = \theta_3\}$ is a subspace of A. Let $P_{\bar{x}}$ be the projection operator with $P_{\bar{x}}A = D_{\bar{x}}$.

---

[1]For finite dimensional spaces, this is equivalent to the rank condition.

[2]This is Gapushkin's (1964) notation.

0.4) Let Z be a linear space, we denote by Z^* the space of line are functionals defined on Z.

0.5) Denote by N_δ the set $N_\delta = \bigcup_{x_0 \in N} \gamma(x_0, \xi)$.

We now list the conditions which we use in the statements of sufficiency theorems for regularity.

(S.1) The function $h'(x, \xi)$ and $g'(x, \xi)$ are continuous and bounded on N.

(S.2) $g'(x, \xi)$ maps A onto C. Furthermore, the space A may be written as the direct sum of $D_{\bar{x}}$ and another subspace $E_{\bar{x}}$, i.e., $A = D_{\bar{x}} \oplus E_{\bar{x}}$ where the projection operator (see 0.3) $P_{\bar{x}}$ is bounded, i.e., there exists a positive constant M such that $\|P_{\bar{x}}\|_A \leq M$.

(S.2)' The space A^* may be written as the direct sum of two subspaces $R_{\bar{x}}^*$ and $S_{\bar{x}}^*$, i.e., $A^* = R_{\bar{x}}^* \oplus S_{\bar{x}}^*$, where $R_{\bar{x}}^* = \{g'^*(\bar{x}, \xi_1^*) = 0, \xi_1^* \in C^*\}$ and where g'^* is the conjugate[1] operator of g' and where $S_{\bar{x}}^*$ is a subspace of A^*. Furthermore, the projection operator $Q_{\bar{x}}$ with $Q_{\bar{x}}A^* = R_{\bar{x}}^*$ is bounded.

(S.3) Given $\delta > 0$. For any $x \in \gamma(\bar{x}, \delta) \cap N$, the space A can be written as the direct sum of $A = D_x \oplus E_x$ and the projection operators P_x with $P_x A^* = R_x$ are bounded and satisfy the Lipschlitz condition, i.e., $\|P_x\| \leq M$ and $\|P_{x_1} - P_{x_2}\|_A \leq M_1 \|X_1 - X_2\|_A$ for $x, x_1, x_2 \in \gamma(\bar{x}, \delta) \cap N$, where M and M_1 are positive constants.

(S.4) $\|g'^*(\bar{x}, \xi_1^*)\| \geq M \|\xi_1^*\|$, for any $\xi_1^* \in C^*$, where $M > 0$.

(S.4)' $\|g'^*(\bar{x}, \eta)\| \geq M_1 \|\eta\|_A$, for any $\eta \in E_{\bar{x}}$, where $M_1 > 0$.

(S.4)" For any $y \in C$, the equation $g'(\bar{x}, b) = y$ has a solution $b(y)$ with $\|b(y)\|_A \leq M_2 \|y\|_C$ where M_2 is a positive constant.

(S.5) There exists $\xi \in A$ with $\|\xi\|_A \leq K$ such that: (i) $g'(x, \xi) = \theta_3$. (ii) $[L, (h(\bar{x}) + h'(\bar{x}, \xi))] \geqq P$, where L is a non-negative linear

---

[1] See Kantorovich (1959) Chapter XII, page 476.

functional with $||L|| = 1$ and where P and K are positive constants.

We now present sufficiency conditions for regularity. These lemmas follows from Gapushkin's theorems (1964) on uniform regularity of N.

Lemma 1. *For equality constraints,* (S.3) $\Rightarrow$ (R.1).

The lemma follows from Theorem 2 of Gapushkin (1964).

Lemma 2. *If* A *is reflexive then* (S.3)' $\Rightarrow$ (R.1) *for equality constraints.*

This follows from lemma 1 (as corollary to theorem 2 of Gapushkin (1964).

Lemma 3. S.1, S.2, *and either* S.4, S.4', *or* S.4" $\Rightarrow$ (R.1) *for equality constraints.*

This follows from theorem 3 of Gapushkin (1964) and its corollaries.

Lemma 4. *In the presence of equality and inequality constraints* any *of the following conditions is sufficient for* (R.3):

(i) *The equality constraint satisfies* (R.1), *and* (S.1) *and* (S.5) *are satisfied.*

(ii) (S.1), (S.3), and (S.5)

(iii) A *is reflexive,* (S.1), (S.3)', *and* (S.5)

(iv) (S.1), (S.2), (S.4)', *and* (S.5)

(v) (S.1), (S.2), (S.4)', *and* (S.5)

(vi) (S.1), (S.2), (S.4)", *and* (S.5).

The lemma follows from theorem 4 of Gapushkin (1964) and from his remark at the end of section 4 of (1964).

4.1.3 FIRST ORDER NECESSARY CONDITIONS FOR THE REGULAR CASE

Theorem 13. *If, in addition to the assumptions of theorem 12,* $\hat{x}$ *is (R.3)-regular then the conclusions of theorem 12 follow with* $\lambda_0 > 0$ *and* ℓ^1 *and* ℓ^2 *are unique (taking* $\lambda_0 = 1$*).*

For the case of equality constraints, the theorem was proved directly[1] by Goldstine (1940), utilizing (S.2) without assuming that A is the direct sum of S.2.[2] The theorem was proved directly by Hurwicz[3] (1958), and it follows from theorem 5 of Gapushkin, who restricts A and C to be reflexive.

4.2 FIRST ORDER SUFFICIENT CONDITIONS

Theorem 14. *If* 1) *The functions* f, g, *and* h *are differentiable,* 2) *The conclusions of theorem 12 are satisfied, with* $\lambda_0 > 0$, *at a point* $\hat{x}$ *that satisfies the constraints of problem 3, and if either:* 3.a) *The functional* F *of theorem 12 is concave,* 3.b) *The equality constraint* g *is linear and* f *and* h *are concave. Then* $\hat{x}$ *is a global solution of problem 3.*

For an outline of the proof of this theorem see the proof of theorem V.3.3. of Hurwicz (1958) where he utilizes the fact that the difference between the values of a concave functional, say J(x), at two different points is less or equal to the differential, i.e., $J(x'') - J(x') \leq J'(x'' - x'))$. Guignard (1969), using a constraint qualification, proves theorem 14 with pseudo-concavity of f and h replacing assumption 3 of the theorem (in the absence of equality constraints).

---

[1] Without using theorem 12, theorem 2.1.

[2] See Liusternik and Sobolev (1951), (page 204) for a proof that this part of (S.2) is dispensable and for an elegant proof of theorem 13 for equality constraints.

[3] Theorem V.3.3.2 (page 97), see remark 1 in 4.1.1 of this paper.

4.3 SECOND ORDER NECESSARY CONDITIONS

Conjecture 1. *If* 1) *the functions* f, g, *and* h *have second order differentials.* 2) $\hat{x}$ *is a solution to problem 3 and* 3) $\hat{x}$ *is regular. Then the upper bound of* $F''(\hat{x}, \xi)$ *is non-positive, for* ξ *with* $||\xi||_A = 1$ *that satisfy* a) $g'(\hat{x}, \xi) = 0$, b) *If* h *is effective i.e., if* $h(\hat{x} = 0$ *then* $h'(\hat{x}, \xi) = 0$, *where* F *is as defined in theorem 12.*

For equality constraints the conjecture was proved by Goldstine[1] (1940).

4.4 SECOND ORDER SUFFICIENT CONDITIONS

Conjecture 2. *If* 1) f, g, *and* h *have second differentials,* 2) *The conclusions of theorem 12 are satisfied at a point* $\hat{x}$ *that satisfies constraints 1) and 2) of problem 3,* 3) *The point* $\hat{x}$ *is regular and* 4) *The upper bound of* $F''(\hat{x}, \xi)$ *is negative for* ξ *with* $||\xi|| = 1$ *satisfying* a) $g'(\hat{x}, \xi) = 0$ *and* b) *if* h *is effective at* $\hat{x}$ *then* $h'(\hat{x}, \xi) = 0$, *where* F *is as defined in theorem 12. Then there exists a neighborhood* $\bar{N}$ *in* A *such that* $f(\hat{x}) \geq f(x)$ *for* $x \in \bar{N} \cap N$.

This conjecture was proved by Goldstine[2] (1940) for the case of equality constraints.

[1]Theorem 2.3 (page 147).
[2]Theorem 3.1 (page 148).

REFERENCES

Abraham-Frois, G. and Berrebi, E. (1976), *Theory of Value, Prices and Accumulation.* Translated from the French edition by M. P. Kregel-Javaux (1979), Cambridge University Press, Cambridge.

Akhiezer, N. I. (1962), *The Calculus of Variations.* English Translation: Blaisdel Publishing Company, New York.

Arrow, K. and Enthoven, A. (1961), Quasi-Concave Programming. *Econometrica* 29: pp. 779-800.

Arrow, K. and Hurwicz, L. (1955) Reduction of Constrained Maxima to Saddle Point Problems. *Proceedings of the Third Berkeley Symposium on Mathematical Statistics and Probability.* University of California Press.

Arrow, K. and Hurwicz, L. (1960), Decentralization and Computation in Resource Allocation. In R. W. Pfouts (ed.) *Essays in Economics and Econometrics.* University of North Carolina Press, Chapel Hill, North Carolina.

Arrow, K. and Hurwicz, L. and Uzawa, H. (1961), Constraints Qualification in Maximization Problems. *Naval Research Logistics Quarterly,* Vol. 8: pp. 175-191.

Averbukh, V. I. and Smolyanov, O. G. (1967), The Theory of Differentiation in Linear Topological Spaces. *Usekhi Mt. Nauk* 22: pp. 201-260; English Translation in *Russian Math. Surveys* (1967) 22: pp. 201-258.

Averbukh, V. I. and Smolyanov, O. G. (1968), The Various Definitions of the Derivative in Linear Topolocial Spaces. *Uspekhi Mt. Nauk* 23: pp. 67-116. English Translation in *Russian Math. Surveys* (1968) 23: pp. 67-113.

Becker, G. (1980), *Human Capital.* University of Chicago Press, Chicago, Illinois.

Beltrami, G. (1987), *Mathematics for Dynamic Modelling.* Academic Press, New York.

Berge, C. (1959), *Espaces Topologique, fonctions multivoques.* Dunod, Paris. English Translation: *Topological Spaces* (1963), MacMillan, New York.

Berkovitz, L. (1963), Variational Problems of Control and Programming. *Journal of Mathematical Analysis and Applications* 3: pp. 145-169.

Bliss, G. (1930), The Problem of Lagrange in the Calculus of Variations. *American Journal of Mathematics* 52: pp. 673-743.

Bliss, G. (1938), Normality and Abnormality in the Calculus of Variations. *Transactions of the American Mathematical Society,* Vol. XLIII: pp. 365-376.

Bliss, G. (1946), *Lectures on the Calculus of Variations.* University of Chicago Press, Chicago.

Blum, E. (1965), Minimization of Functionals with Equality Constraints. *SIAM Journal on Control,* Vol. 3: pp. 299-317.

Bohm-Bawerk, E. (1870), *Capital and Interest.* MacMillan, London.

Brady, C. (1938), The Minimum of a Function of Integrals in the Calculus of Variations. *Contributions to the Calculus of Variations 1938-1941*, University of Chicago Press, Chicago.

Brown, D. and Heal, G. (1979), Equity, Efficiency and Increasing Returns. *Review of Economic Studies.* Vol. XLIV (4), No. 145: pp. 571-586.

Burger, E. (1955), On Extrema with Side Conditions, *Econometrica*, Vol. 23, pp. 451-452.

Canon, M., Cullum, C., and Polak, E. (1970), *Theory of Optimal Control and Mathematical Programming.* McGraw-Hill Book Company, New York.

Caratheodory, C. (1935), *Variationsrechnung und partielle Differentialgleichungen erster Ordnung.* B. G. Teubner, Berlin. English Translation: *Calculus of Variations and Partial Differential Equations of the First Order,* Vols. I and II, (1965 and 1967) Holen-Day, San Fransisco, California.

Cesari, L. (1965), An Existence Theorem in Problems of Optimal Control. *SIAM Journal on Control,* Vol. 3: pp. 7-22.

Debreu, G. (1952), Definite and semidefinite Quadratic Forms, *Econometrica*, Vol. 20, pp. 295-300.

Dubovskii, A. Ya. and Milyutin, A. A. (1965), Extremum Problems in the Presence of Restrictions. *Zh. Vychisl Fiz.* 5: pp. 395-453. English Translation: *U.S.S.R. Computational Mathematics and Math Physics* (1965) 5: pp. 1-80.

El'sgolts, L. E. (1955), *Qualitative Methods in Mathematical Analysis.* Moscow. English Translation as Vol. 12, *Translations of Mathematical Monographs* (1964), American Mathematical Society. Providence, Rhode Island.

El-Hodiri, M. (1969), Notes on Optimization Theory, Part I: Finite Dimensional Problems. *Research Papers in Theoretical and Applied Economics,* No. 21, University of Kansas.

El-Hodiri, M. (1970), Constrained Extrema, Introduction to the Differentiable Case with Economic Applications. No. 56 Lecture Notes in *Operations Research and Mathematical Systems.* Springer-Verlag, Berlin, Heidelberg, New York.

Evans, G. C. (1924), The Dynamics of Monopoly. *American Math. Monthly* 31: pp. 77-83.

Evans, G. C. (1930), *Mathematical Introduction to Economics.* McGraw-Hill, London.

Ewing, G. (1969), *Calculus of Variations with Applications.* Norton and Co., New York.

Fenchel, W. (1953), *Convex Cones, Sets, and Functions.* Lecture notes, Princeton University, Department of Mathematics, New Jersey.

Fleming, W. (1977), *Functions of Several Variables* (2 ed.). Springer-Verlag, New York, Heidelberg.

Gantmacher, F. R. (1959), *The Theory of Matrices.* Chelsea, New York.

Gapushkin, V. F. (1964), On Critical Points of Functionals in Banach Spaces. *Mat. Sbornik* 64: pp. 589-617. English Translation in *Three Papers on Optimization Theory,* PEREC Report (1966). Purdue University, Dept. of Electrical Engineering.

Gelfand, I. M. and Fomin, S. V. (1963), *Calculus of Variations* (R. Silverman, translator from Russian). Prentice Hall, Englewood Cliffs, New Jersey.

Gibson, C. (1979), *Singular Points of Smooth Mappings*. Pitman, San Francisco, California.

Goldstine, Herman H. (1940), Minimum Problems in the Functional Calculus. *Bul. American Math. Society* 46: pp. 142-149.

Golubitsky, M. and Guillemin, B. (1972), *Stable mappings and their Singularities*. Springer-Verlag, New York.

Guignard, M. (1969), Generalized Kuhn-Tucker conditions for Mathematical Programming Problems in a Banach Space. *SIAM Journal on Control* 7: pp. 232-241.

Halanay, A. (1968), Optimal Control Systems with Time Lag, *SIAM Journal on Control* 6: pp. 215-234.

Hestenes, M. (1965), On Variational Theory and Optimal Control Theory. *SIAM Journal on Control* 3: pp. 23-48.

Hestenes, M. (1966), *Calculus of Variations and Optimal Control Theory*. John Wiley & Sons, New York.

Hestenes, M. (1975), *Optimization Theory*. John Wiley & Sons, New York.

Hurwicz, L. (1958), Programming in Linear Spaces. In Arrow, Hurwicz and Uzawa (eds). *Studies in Linear and Nonlinear Programming*. Stanford University Press, Stanford, California.

Ioffe, A. D. and Tikhomirov, V. M. (1974), *Theory of Extremal Problems*. Nauka, Moscow. English Translation (1979), North Holland, New York.

John, Fritz. (1948), Extremum Problems with Inequalities and Subsidiary Conditions. In K. O. Friedrick, O. E. Neugebaur and J. J. Stokes (eds). *Studies and Essays: Courant Anniversary Volume*. Interscience Publishers, New York, New York.

Kannai, Y. (1977), Concavifiability and Constructions of Concave Utility Functions. *Journal of Mathematical Economics* 4: pp. 1-56.

Kantorovich, L. V. and Akilov, G. P. (1959), *Functional Analysis in Normal Linear Spaces.* Fizgmatiz, Moscow. English Translation (1964): Pergamon Press.

Karush, W. (1939), *Minima of Functions of Several Variables with Inequalities as Side Conditions.* University of Chicago Master's Thesis.

Kuhn, H. (1976), Nonlinear Programming: A Historical View. *Nonlinear Programming.* American Math. Society, Providence, Rhode Island.

Kuhn, H. W. and Tucker, A. W. (1951), Nonlinear Programming. In *Proc. Second Berkeley Symposium on Math. Stat. and Probability,* J. Neuman (ed.). University of California Press, Berkeley, California.

Lagrange, L. (1762), (Actual name is Louis de la Grange) Recherches sur la Methode de Maximis et minimis. *Recuelles de L'Académie de Turin, Miscellanea Taurinesea.* Tom I.

Lagrange, L. (1762), Essai Sur une Nouvelle Methode Pour Determiner les Maxima et Minima des Formules Integrales Indefinies, *Miscellanea Taurinesia,* Tom II. pp. 173-195.

Lee, E. and Markus, L. (1967), *Foundations of Optimal Control Theory.* John Wiley & Sons, New York, New York.

Liusternik, L. and Sobolev, V. (1951), *Elements of Functional Analysis.* Nauka, Moscow. English Translation (1961): Ungar Publishing Co., New York, New York.

Luenberger, D. (1968), Quasi-Convex Programming. *SIAM Journal of Applied Mathematics* 16: pp. 1090-1095.

Mangasarian, O. L. (1966), Sufficient Conditions for the Optimal Control of Nonlinear Systems. *SIAM Journal on Control* 4: pp. 139-152.

McShane, E. (1939), On Multipliers for Lagrange Problems. *American Journal of Mathematics* 61: pp. 809-819.

McShane, E. (1941), Sobre la teoria de los estremos relativos. *Revista De Ciencias, Lima,* Vol. 43: pp. 1111-134, 475-482, 629-666.

McShane, E. (1940), Existence Theorems for the Problem of Balza in the Calculus of Variations. *Duke Mathematical Journal* 7: pp. 28-61.

McShane, E. J. (1973), The Lagrange Multiplier Rule. *American Math. Monthly*, 80: pp. 923-925.

McShane, E. and Botts, A. (1954), *Real Analysis*. D. Van Nostrand, Princeton, New Jersey.

Munkres, J. R. (1968), *Elementary Differential Topology*. Princeton University Press, Princeton, New Jersey.

Neustadt, L. (1967), An Abstract Variational Theory with Applications to a Broad Class of Variational Problems, I and II. *SIAM Journal on Control* 4: pp. 505-528 and 5: 90-138.

Pars, L. (1962), An Introduction to the Calculus of Variations. John Wiley & Sons, New York, New York.

Pennisi, L. L. (1953), An Indirect Sufficiency Proof for the Problem of Lagrange with Differential Inequalities as Added Side Conditions. *Transactions of the American Math. Society* 74: pp. 177-198.

Pontriagin, L. S., Boltyanskii, Gamkrelidze, and Mishchenko (1961) *The Mathematical Theory of Optimal Processes*. Fizgmatiz, Moscow. English Translation (1962), John Wiley & Sons, New York, New York.

Qayum, A. (1963). Investment Maximizing Tax Policy. *Public Finance*, pp. 3-4.

Roberts, A. W. and Varberg, D. E. (1973), *Convex Functions*. Academic Press, New York.

Roemer, J. (1980), A General Equilibrium Approach to Marxian Economics. *Econometrica* 48: p. 505.

Roemer, J. (1982), *A General Theory of Exploitation and Class*. Harvard University Press, Cambridge, Massachusetts.

Roemer, J. (1986), *Value Exploitation, and Class*. Harood, New York.

Russak, I. (1970), On Problems with Bounded State Variables. *Journal of Optimization Theory and Applications* 5: pp. 114-157.

Saaty, T. L. and Bram, J. (1964), *Nonlinear Mathematics*, McGraw-Hill, New York.

Samuelson, P. A. (1947), *Foundations of Economic Analysis*. Harvard University Press, Cambridge, Massachusetts.

Samuelson, P. A. (1959), A Modern Treatment of the Ricardian Economy: The Pricing of Goods and of Labor and Land Services. *Quarterly Journal of Economics* pp. 1-35.

Shostak, R. Ya. (1954), On a Test for Definiteness of Quadratic Forms under Linear Constraints and a Sufficiency Test for Constrained Extrema of Functions of Several Variables (in Russian). *Uspekhi Mat Nauk,* 9(60): pp. 199-206.

Simon, C., Scalar and Vector Maximization: Calculus Techniques with Economic Applications, *Studies in Mathematical Economics,* Stanley Reiter (ed), The Mathematical Association of America, USA.
Smale, S. (1974), Global Analysis and Economics II. A. *Journal of Mathematical Economics.* Vol. 1: pp. 1-14.

Takayama, A. *Mathematical Economics.* The Dryden Press, Hinsdale, Illinois.
Tikhomirov, V. M. (1986), *Fundamental Principles of the Theory of Extremal Problems.* John Wiley & Sons, New York.

Vainberg, M. M. (1956), *Variational Methods for the Study of Nonlinear Operators* Godtekhizdat, Moscow. English Translation (1964): Holden-Day, San Fransisco, California.

Valentine, Fredric (1937), The Problem of Lagrange with Differential Inequalities as Added Side Condition. *Contributions to the Calculus of Variations 1933-1937.* University of Chicago Press, Chicago, Illinois.

Wiggins, S. (1988), *Global Bifurcations and Chaos,* Springer-Verlag, New York.

Young, L. C. (1969), *Lectures on the Calculus of Variations* and *Optimal Control Theory.* W. B. Saunders and Co., Philadelphia, Pennsylvania.

Young, W. H. (1909), *The Fundamental Theorems of the Differential Calculus,* Hefner, New York.

EPILOGUE

"Thus spoke uncle Mahmoud Bagara'a to Myrtyanu:

My son
Bear with me five more minutes
I will be done and you gone."

Credo of Sheik Seleem College
Munsa, Ashmoon, Minufia